Firefighter Fatalities in the United States in 2008

Prepared by

U.S. Department of Homeland Security

Federal Emergency Management Agency

U.S. Fire Administration

National Fire Data Center

and

The National Fallen Firefighters Foundation
www.firehero.org

- - -

In memory of all firefighters
who answered their last call in 2008
To their families and friends
To their service and sacrifice

Cover photo courtesy of Jon Vait.

Table of Contents

Photographers contributing photos to this report:

Jon Vait, Eagle Point, OR

Jayson Coil, Sedona Fire District, AZ (www.jaysoncoil.com)

Brandan Schulze, Olympic National Forest, WA

Acknowledgements

This study of firefighter fatalities would not have been possible without the cooperation and assistance of many members of the fire service across the United States. Members of individual fire departments, chief fire officers, wildland fire service organizations such as the United States Forest Service (USFS), the National Park Service (NPS), the Bureau of Land Management (BLM), the Bureau of Indian Affairs (BIA), the U.S. Fish and Wildlife Service (FWS), as well as the U.S. Department of Justice (DOJ), the National Fire Protection Association (NFPA), and many others contributed important information for this report.

The National Fallen Firefighters Foundation (NFFF) was responsible for compilation of a large portion of the data used in this report and the incident narrative summaries found in Appendix A.

The ultimate objective of this effort is to reduce the number of firefighter deaths through an increased awareness and understanding of their causes and how they can be prevented. Firefighting, rescue, and other types of emergency operations are essential activities in an inherently dangerous profession, and unfortunate tragedies do occur. This is the risk all firefighters accept every time they respond to an emergency incident. However, the risk can be greatly reduced through efforts to improve training, emergency scene operations, and firefighter health and safety initiatives.

Background

For 32 years, the U.S. Fire Administration (USFA) has tracked the number of firefighter fatalities and conducted an annual analysis. Through the collection of information on the causes of firefighter deaths, the USFA is able to focus on specific problems and direct efforts toward finding solutions to reduce the number of firefighter fatalities in the future. This information is also used to measure the effectiveness of current programs directed toward firefighter health and safety.

Several programs have been funded by USFA in response to this detailing of firefighter fatalities. For example, USFA has sponsored significant work in the areas of general emergency vehicle operations safety, fire department tanker/tender operations safety, firefighter incident scene rehabilitation, and roadside incident safety. The data developed for this report are also widely used in other firefighter fatality prevention efforts.

One of USFA's main program goals is a 25-percent reduction in firefighter fatalities in 5 years and a 50-percent reduction within 10 years. The emphasis placed on these goals by USFA is underscored by the fact that these goals represent one of the five major objectives that guide the actions of USFA.

In addition to the analysis, USFA, working in partnership with the NFFF, develops a list of all onduty firefighter fatalities and associated documentation each year. If certain criteria are met, the fallen firefighter's next of kin, as well as members of the individual's fire department, are invited to the annual Fallen Firefighters Memorial Service. The service is held at the National Emergency Training Center (NETC) in Emmitsburg, Maryland, during Fire Prevention Week in October of each year. Additional information regarding the Memorial Service can be found at www.firehero.org or by calling the NFFF at (301) 447-1365.

Other resources and information regarding firefighter fatalities, including current fatality notices, the National Fallen Firefighters Memorial database, and links to the Public Safety Officers' Benefit (PSOB) Program can be found at www.usfa.dhs.gov/fireservice/fatalities/

Introduction

This report continues a series of annual studies by USFA of onduty firefighter fatalities in the United States.

The specific objective of this study is to identify all onduty firefighter fatalities that occurred in the United States and its protectorates in 2008 and to analyze the circumstances surrounding each occurrence. The study is intended to help identify approaches that could reduce the number of firefighter deaths in future years.

Who is a Firefighter?

For the purpose of this study, the term firefighter covers all members of organized fire departments with assigned fire suppression duties in all 50 States, the District of Columbia, and the Territories of Puerto Rico, the Virgin Islands, American Samoa, the Commonwealth of the Northern Mariana Islands, and Guam. It includes career and volunteer firefighters; full-time public safety officers acting as firefighters; fire police; State, territory, and Federal government fire service personnel, including wildland firefighters; and privately employed firefighters, including employees of contract fire departments and trained members of industrial fire brigades, whether full- or part-time. It also includes contract personnel working as firefighters or assigned to work in direct support of fire service organizations (air-tanker crews).

Under this definition, the study includes not only local and municipal firefighters, but also seasonal and full-time employees of USFS, BLM, BIA, U.S. FWS, NPS, and State wildland agencies. The definition also includes prison inmates serving on firefighting crews; firefighters employed by other governmental agencies, such as the U.S. Department of Energy (DOE); military personnel performing assigned fire suppression activities; and civilian firefighters working at military installations.

What Constitutes an Onduty Fatality?

Onduty fatalities include any injury or illness sustained while on duty that proves fatal. The term "onduty" refers to being involved in operations at the scene of an emergency, whether it is a fire or nonfire incident; responding to or returning from an incident; performing other officially assigned duties such as training, maintenance, public education, inspection, investigations, court testimony, and fundraising; and being on-call, under orders, or on standby duty except at the individual's home or place of business. An individual who experiences a heart attack or other fatal injury at home while he or she prepares to respond to an emergency is considered onduty when the response begins. A firefighter that becomes ill while performing fire department duties and suffers a heart attack shortly after arriving home or at another location may be considered onduty since the inception of the heart attack occurred while the firefighter was on duty.

On December 15, 2003, the President of the United States signed into law the Hometown Heroes Survivors Benefit Act of 2003. After being signed by the President, the Act became Public Law 108-182. The law presumes that a heart attack or stroke are in the line of duty if the firefighter was engaged in nonroutine stressful or strenuous physical activity while on duty and the firefighter becomes ill while on duty or within 24 hours after engaging in such activity. The full text of the law is available at:
http://frwebgate.access.gpo.gov/cgi-bin/getdoc.cgi?dbname=108_cong_publiclaws&docid=f:publ182.108.pdf

The inclusion criteria for this study have been impacted by this change in the law. Previous to December 15, 2003, firefighters who became ill as the result of a heart attack or stroke after going off duty needed to register some complaint of not feeling well while still on duty in order to be included in this study. For firefighter fatalities after December 15, 2003, firefighters will be included in this study if they become ill as the result of a heart attack or stroke within 24 hours of a training activity or emergency response. Firefighters who become ill after going off duty where the activities while on duty were limited to tasks that did not involve physical or mental stress will not be included in this study.

A fatality may be caused directly by an accidental or intentional injury in either emergency or nonemergency circumstances, or it may be attributed to an occupationally related fatal illness. A common example of a fatal illness incurred on duty is a heart attack. Fatalities attributed to occupational illnesses would also include a communicable disease contracted while on duty that proved fatal when the disease could be attributed to a documented occupational exposure.

Firefighter fatalities are included in this report even when death is considerably delayed after the original incident. When the incident and the death occur in different years, the analysis counts the fatality as having occurred in the year in which the incident took place. One firefighter died in 2008 from injuries sustained in a 1999 incident, bringing that year's total to 114. Information about this death in Massachusetts is included in Appendix A of this report.

There is no established mechanism for identifying fatalities that result from illnesses such as cancer that develop over long periods of time and which may be related to occupational exposure to hazardous materials or toxic products of combustion. It has proved to be very difficult over the years to provide a complete evaluation of an occupational illness as a causal factor in firefighter deaths due to the following limitations: the exposure of firefighters to toxic hazards is not sufficiently tracked; the often delayed long-term effects of such toxic hazard exposures; and the exposures firefighters may receive while off duty.

Sources of Initial Notification

As an integral part of its ongoing program to collect and analyze fire data, the USFA solicits information on firefighter fatalities directly from the fire service and from a wide range of other sources. These sources include the PSOB Program administered by the DOJ, the National Institute for Occupational Safety and Health (NIOSH), the Occupational Safety and Health Administration (OSHA), the Department of Defense (DOD), the National Interagency Fire Center (NIFC), and other Federal agencies.

The USFA receives notification of some deaths directly from fire departments, as well as from such fire service organizations as the International Association of Fire Chiefs (IAFC), the International Association of Fire Fighters (IAFF), the NFPA, the National Volunteer Fire Council (NVFC), State fire marshals, State fire training organizations, other State and local organizations, fire service Internet sites, news services, and fire service publications. The USFA also keeps track of fatal fire incidents as part of its Major Fires Investigation Program and performs an ongoing analysis of data from the National Fire Incident Reporting System (NFIRS).

Procedure for Including a Fatality in the Study

In most cases, after notification of a fatal incident, initial telephone contact is made with local authorities by the USFA to verify the incident, its location, jurisdiction, and the fire department or agency involved. Further information about the deceased firefighter and the incident may be obtained from the chief of the fire department or designee over the phone or by other data collection forms. After basic information is collected, a notice of the firefighter fatality is posted at the National Fallen Firefighters Memorial site in Emmitsburg, Maryland, the USFA Web site, and a notice of the fatality is transmitted by electronic mail to a large list of fire service organizations and fire service members.

Information that is requested routinely from fire departments that have experienced a fatality includes NFIRS-1 (incident) and NFIRS-3 (fire service casualty) reports; the fire department's own incident and internal investigation reports; copies of death certificates and autopsy results; special investigative reports; law enforcement reports; photographs and diagrams; and newspaper or media accounts of the incident. Information on the incident may also be gathered from NFPA or NIOSH reports on an incident.

After obtaining this information, a determination is made as to whether the death qualifies as an onduty firefighter fatality according to the previously described criteria. With the exception of firefighter deaths after December 15, 2003, the same criteria were used for this study as in previous annual studies. Additional information may be requested by USFA, either through followup with the fire department directly, from State vital records offices, or other agencies. The final determination as to whether a fatality qualifies as an onduty death for inclusion in this statistical analysis is made by the USFA. The final determination as to whether a fatality qualifies as a line-of-duty death (LODD) for inclusion in the annual Fallen Firefighters Memorial Service is made by NFFF.

2008 Findings

One-hundred and eighteen (118) firefighters died while on duty in 2008, the same number of firefighter fatalities as the previous year. This total includes firefighters who died under circumstances that are included in this report as a result of inclusion criteria changes resulting from the Hometown Heroes Act of 2003.

The number of firefighter fatalities using the pre-Hometown Heroes criteria was 107, up by two fatalities in comparison to 2007.

An analysis of multiyear firefighter fatality trends needs to acknowledge the changes from the Hometown Heroes Survivors Benefit Act of 2003. While a change in reporting criteria stemming from the Act does not diminish the sacrifices made by any firefighter who dies while on duty, or the sacrifices made by his/her family and peers, some graphs and charts will indicate the Hometown Heroes portion of the total.

Moreover, when conducting multiyear comparisons of firefighter fatalities in this report, the losses that were the result of the attacks on the World Trade Center (WTC) in New York City on September 11, 2001, are sometimes also set apart for illustrative purposes. This action is by no means a minimization of the supreme sacrifice made by these firefighters.

Photo courtesy of Jayson Coil.

Figure 1. Onduty Firefighter Fatalities (1977-2008)

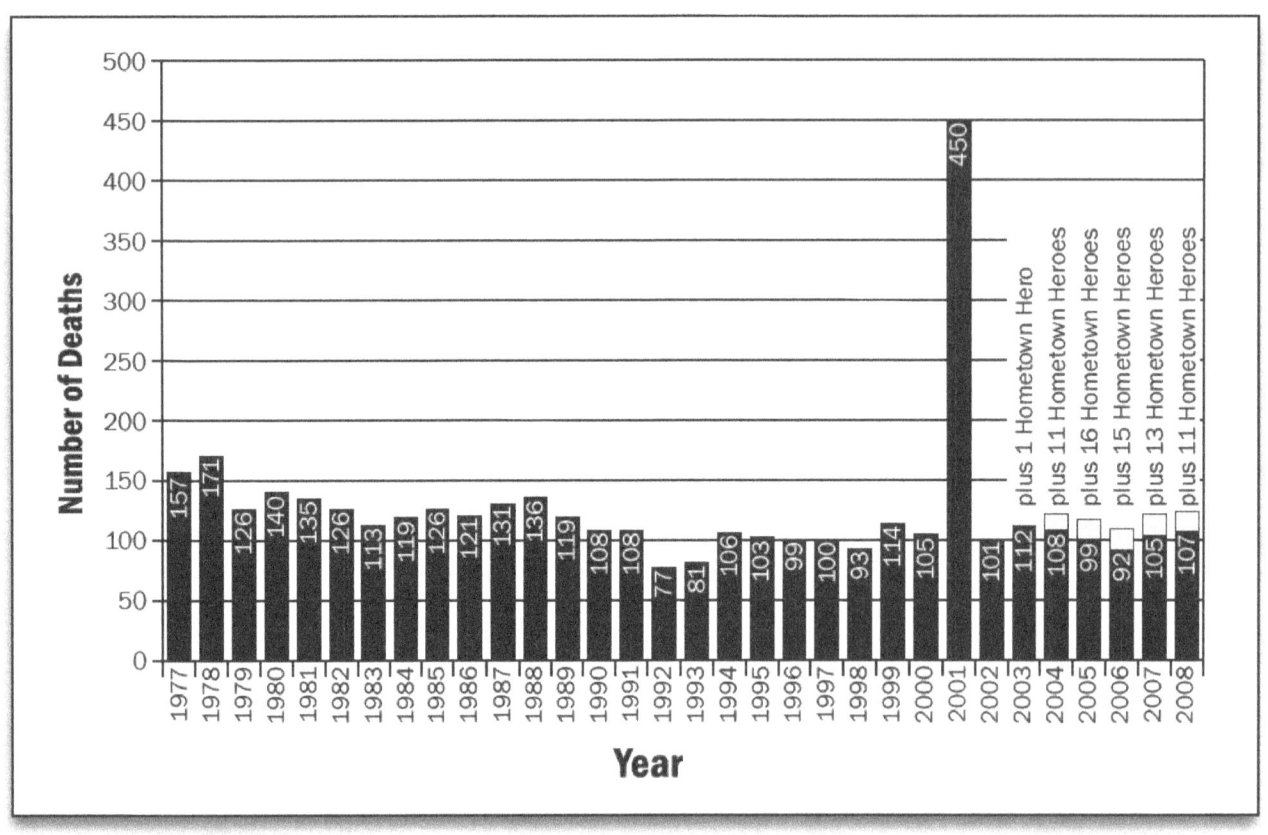

Figure 2. Firefighter Fatalities per 100,000 Fires

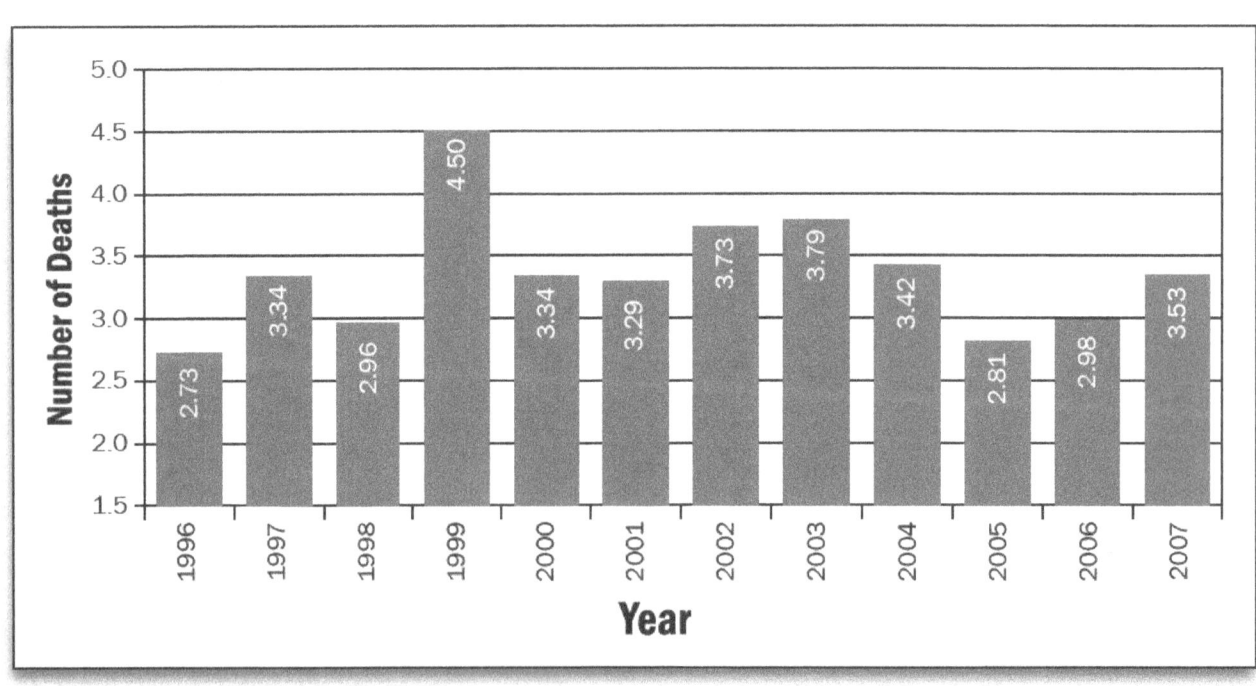

Career, Volunteer, and Wildland Agency Deaths

In 2008, firefighter fatalities included 66 volunteer firefighters, 34 career firefighters, and 18 part-time or full-time members of wildland or wildland contract fire agencies. (Figure 3).

Figure 3. Career, Volunteer, and Wildland Agency Deaths (2008)

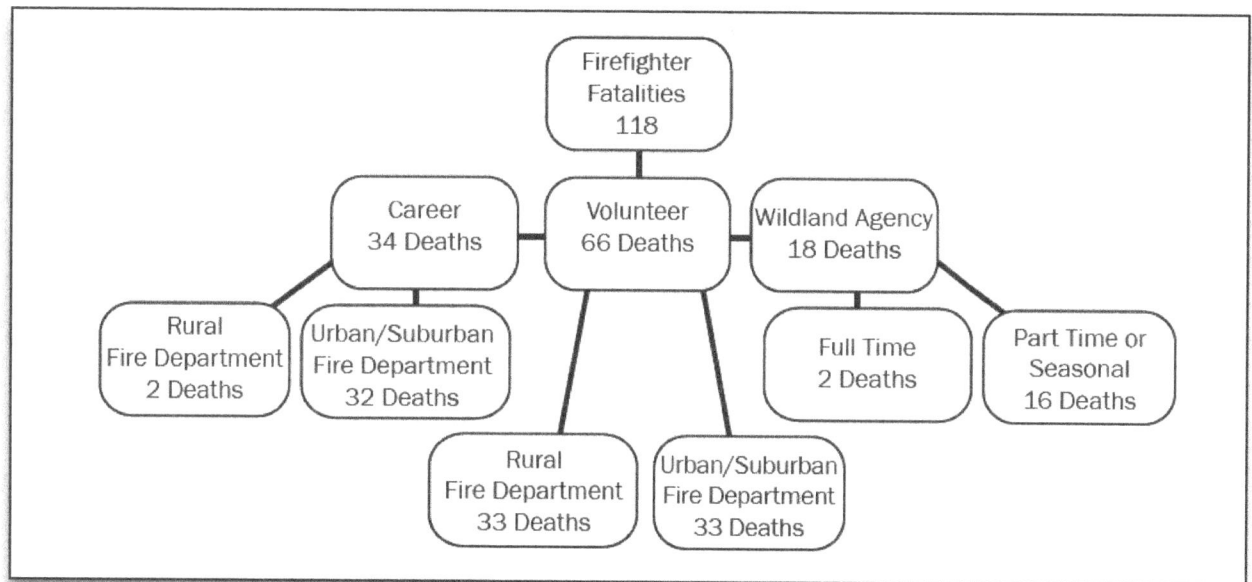

Figure 4. Career, Volunteer, Industrial and Wildland Agency Deaths (1990-2008)

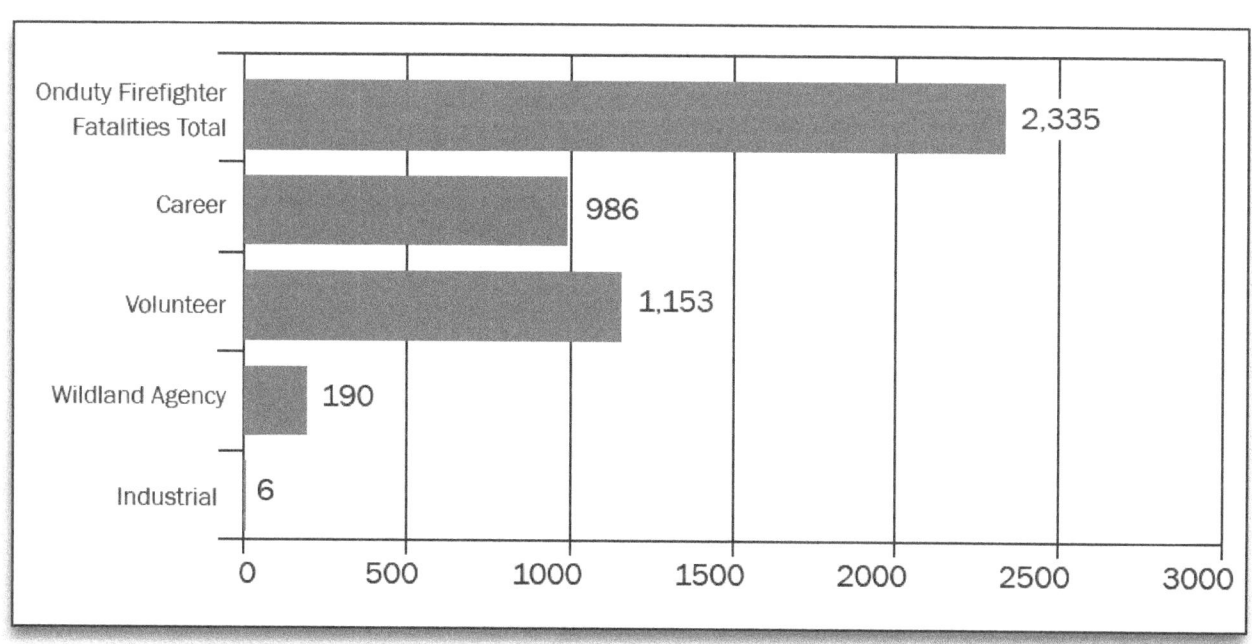

7

Gender

Five of the firefighters who died in 2008 were female and 113 were male.

> Since 1990, 67 of 2,335 firefighters who died while on duty were female and 2,268 were male.

Table 1: Firefighter Fatality by Gender

Year	Death	Gender
1999	111	M
1999	3	F
2000	102	M
2000	3	F
2001	445w/WTC	M
2001	5	F
2002	96	M
2002	5	F
2003	110	M
2003	3	F
2004	113	M
2004	6	F
2005	112	M
2005	3	F
2006	101	M
2006	6	F
2007	115	M
2007	3	F
2008	113	M
2008	5	F

Multiple Firefighter Fatality Incidents

The 118 deaths in 2008 resulted from a total of 105 fatal incidents. There were 5 firefighter fatality incidents where 2 or more firefighters were killed in 2008, claiming a total of 18 firefighters.

Table 2: Multiple Firefighter Fatality Incidents

Year	Number of Incidents	Total Number of Deaths
2008	5	18
2007	7	21
2006	6	17
2005	4	10
2004	3	6
2003	7	20
2002	9	25
2001	8	362
2001 w/o WTC	7	18
2000	5	10
1999	6	22
1998	10	22

- Two firefighters were killed in a large manufacturing occupancy (approximately 70,000 square feet) when they were overcome by fire conditions that changed rapidly due to the collapse of an interior wall.

- Two firefighters were killed when they were caught and trapped by a structural collapse in a residential fire.

- Two firefighters were killed while responding to the Ordway wildland fire when a bridge that had been damaged by fire failed as their brush truck drove over it. The deaths are the only wildland fire-related firefighter fatalities on record related to a structural collapse.

- Nine firefighters, including two pilots assigned to the Iron Complex fire, were killed when their helicopter experienced a loss of power to the main rotor during takeoff and subsequently impacted trees and terrain.

- Three firefighters, two pilots, and the crew chief of an air tanker (Neptune Tanker 09) under contract with the USFS, crashed moments after takeoff from the Reno-Stead Airport. The crew was dispatched to provide a water drop on a wildland fire burning in Calaveras County, California.

> For the years 1990-2008, 28.6 percent of all 2,335 onduty firefighter fatalities occurred as part of a multiple fatality incident. If one were to remove the losses from September 11, 2001, (WTC Terrorist Attacks) and wildland fire-related incidents from the calculation, only 11.6 percent of the remaining 1,638 onduty firefighter fatalities resulted from multiple fatality incidents. Of the 352 firefighter fatalities associated with wildland fire incidents, 38 percent were from multiple fatality incidents.

Wildland Firefighting Deaths

In 2008, 26 firefighters were killed during activities involving brush, grass, or wildland firefighting. This total includes part-time and seasonal wildland firefighters, full-time wildland firefighters, and municipal or volunteer firefighters whose deaths are related to a wildland fire (Figure 5).

Figure 5. Firefighter Fatalities Related to Wildland Firefighting (1997-2008)

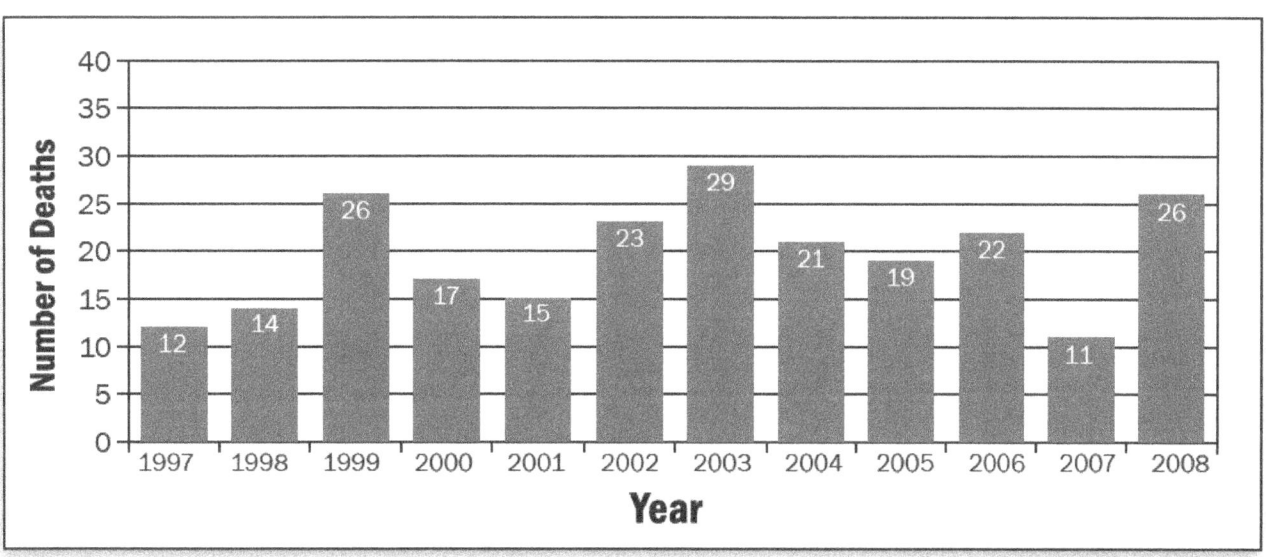

Two firefighters died when their brush truck was involved in a noncollision fall due to structural collapse of a bridge they were crossing that had been undermined by fire.

One firefighter was killed when the Single Engine Air Tanker (SEAT) he was piloting crashed.

One firefighter died of a heart attack while riding in a grass truck responding to an outdoor fire.

One firefighter died of multiple blunt trauma when he was struck by a vehicle entering the scene of a multiple vehicle collision. A contributing cause was heavy smoke from an outdoor fire and fog obscuring vision along the roadway. A sheriff's deputy was also struck and killed and another deputy was injured in the incident.

One firefighter died from a nontraumatic brain hemorrhage several hours after returning with his Strike Team from the scene of a wildland fire.

One firefighter died when the medical helicopter he was being transported in collided with another medical helicopter, killing the firefighter and six others. The firefighter had been battling a fire on the north rim of the Grand Canyon National Park when he was bitten by an insect and taken to a nearby hospital for treatment. While in the hospital, he suffered anaphylactic shock from the antibiotics being used to treat the insect bite and it became necessary for the firefighter to be flown to a larger medical center.

One firefighter assigned the position of lookout on a wildland fire was helping carry hose up a hill when he experienced extreme fatigue and respiratory distress. He was transported to the hospital where he died the following day from a massive heart attack.

One firefighter working tree felling operations was struck and injured by a tree. Due to heavy smoke conditions, the firefighter had to be carried a distance before he could be evacuated by a U.S. Coast Guard (USCG) helicopter. While being transported aboard the helicopter, the firefighter went into cardiac arrest and died.

One firefighter, in preparation to assume management responsibility for a wildland fire, was scouting the area of operations when the fire spread quickly and burned over his position.

Nine firefighters, including two pilots assigned to the Iron Complex fire, were killed when their helicopter experienced a loss of power to the main rotor during takeoff, and subsequently impacted trees and terrain.

One firefighter died from a heart attack a short time after he returned home from fighting lightning-caused wildland fires.

One firefighter died from injuries sustained from a fall while scouting a fire in extremely rough terrain and dangerous rock cliffs.

One firefighter died from injuries sustained when he fell from a piece of heavy machinery while clearing fire breaks.

Three firefighters, two pilots, and the crew chief of an air tanker under contract with the USFS, crashed moments after take-off.

One firefighter collapsed and died from a heart attack while supervising a prison firefighting crew.

Table 3: Wildland Firefighting Aircraft Deaths

Year	Total Number of Deaths	Number of Fatal Incidents
2008	16	4
2007	1	1
2006	8	3
2005	6	2
2004	3	3
2003	7	4
2002	6	3
2001	6	3
2000	6	5
1999	0	0
1998	3	2

In 2008, there were three multiple firefighter fatality incidents related to wildland firefighting that killed 14 firefighters.

Table 4: Firefighter Deaths Associated with Wildland Firefighting

Year	Total Number of Deaths	Number of Fatal Incidents	Number of Firefighters Killed in Multiple-Death Incidents
2008	26	15	14
2007	11	11	0
2006	22	13	13
2005	19	15	6
2004	21	21	0
2003	30	22	10
2002	23	14	13
2001	15	9	9
2000	19	16	6
1999	27	26	2
1998	14	13	2

The loss of nine wildland firefighters in the August 5, 2008, "Iron 44" incident was the second largest multiple wildland firefighter fatality incident in decades. The July 7, 1994, Storm King Mountain incident in Colorado killed 14 firefighters. The National Wildfire Coordinating Group (NWCG) reported that the Big Burn of 1910 in Northern Idaho and Western Montana burned millions of acres and took the lives of at least 78 wildland firefighters.

Type of Duty

Activities related to emergency incidents resulted in the deaths of 75 firefighters in 2008, which is one less death than 2007 (Figure 6). This includes all firefighters who died responding to an emergency or at an emergency scene, returning from the emergency incident, and other emergency-related activities. Nonemergency activities accounted for 43 fatalities. Nonemergency duties include training, administrative activities, performing other functions that are not related to an emergency incident, and postincident fatalities where the firefighter does not experience the illness or injury during the emergency.

Figure 6. Firefighter Deaths by Type of Duty (2008)

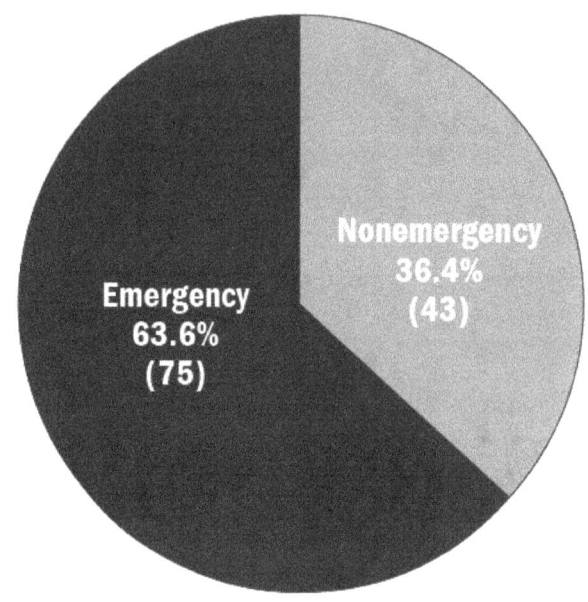

A multiyear historical perspective concerning the percentage of firefighter deaths that occurred during emergency duty is presented in Table 5. The percentage of all deaths for 2004 and the years after are lower for each year due to the inclusion of firefighters covered by the changes resulting from the Hometown Heroes Act of December 2003. As such, a second column has been added to Table 5 for purposes of a longer-term trend analysis.

Table 5: Emergency Duty Firefighter Deaths

Year	Percentage of All Deaths	Percentage of All Deaths Without Hometown Heroes
2008	63.5	70
2007	64.4	72.4
2006	57.5	66.3
2005	52.1	60.6
2004	68.9	75.9
2003	69	69.6
2002	73	73
2001	65	65
2001 with WTC	92	92
2000	71	71
1999	87	87
1998	77	77

The number of deaths by type of duty being performed in 2008 is shown in Table 6 and presented graphically in Figure 7.

Table 6: Firefighter Deaths by Type of Duty (2008)

Type of Duty	Number of Deaths
Fireground Operations	28
Responding/Returning	24
Other Onduty Deaths	30
Training	12
Nonfire Emergencies	11
After an Incident	13
Total	118

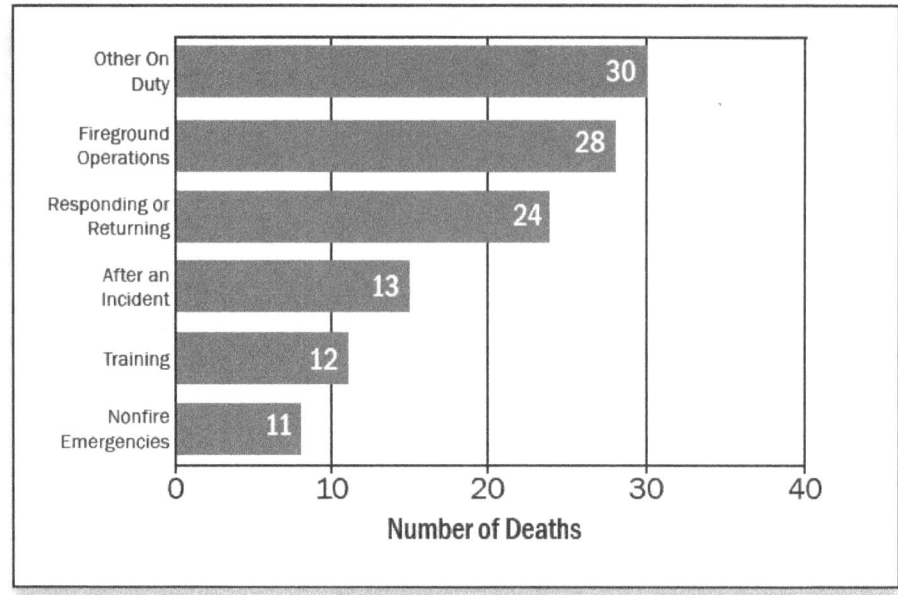
Figure 7: Firefighter Deaths by Type of Duty (2008)

Fireground Operations: Onscene Structure Fire

Of the 28 firefighters killed during fireground operations in 2008, 21 firefighters died while on the scene of a structure fire.

Table 7: Onscene Structure Fire by Cause of Fatal Injury 2004-2008

Cause of Fatal Injury	2004	2005	2006	2007	2008
Caught/Trapped	6	4	4	6	5
Collapse	6	1	8	7	5
Contact with	0	0	0	0	1
Exposure	0	1	0	0	0
Fall	1	2	0	1	1
Lost	0	1	3	11	1
Out of Air	0	0	0	0	1
Stress/Overexertion	7	9	7	9	5
Struck by	1	1	0	1	2
Vehicle Collision	1	0	0	0	0

Over a 5-year period of time, 119 firefighters died while on the scene of a structure fire, 84 in single firefighter fatality incidents and 35 in 14 multiple firefighter fatality incidents.

One firefighter died of smoke inhalation and thermal burns on the 14th floor of a highrise residential occupancy building. The firefighter ran out of air less than 20 minutes after donning his facepiece and was unable to exit the apartment before being overcome.

14

One firefighter was electrocuted while on the platform of a tower ladder as he ascended through powerlines to position the ladder in a spot that would allow access to the upper floors of a residence. A utility company worker had arrived at the scene and departed before the fatal incident took place.

One firefighter died from burns when fire intensified and trapped him on the second floor of a duplex. He was conducting a search for a fire victim that had been confirmed to still be in the residence.

One firefighter was trapped in the basement of a residential structure while attempting to exit the structure after controlling all visible fire but encountering extreme heat conditions.

Two firefighters were trapped as fire conditions rapidly changed due to the collapse of an interior wall of a large 79,000 square foot manufacturing occupancy.

One firefighter died from a heart attack after advancing a hoseline into a residential structure to extinguish a kitchen fire.

One firefighter was wetting down hot spots in the interior of a vacant residential structure when he collapsed and died from a heart attack.

One firefighter died from injuries sustained when he was assigned to force entry into a room that housed electrical equipment and had been emitting smoke. The rotary saw used to attempt to gain access caused sparks that ignited combustible gases that had accumulated in the locked room which resulted in a large explosion.

Two firefighters died from smoke inhalation when they became trapped while operating a hoseline in a residential structure and a large portion of the first floor collapsed.

One firefighter died from burns sustained when overcome by rapid fire progress while working a hoseline inside of a residential structure.

One firefighter in command of an exterior fire at a plastics manufacturing facility was struck and killed by a motorized water monitor; 30 feet of aluminum pipe projected out of the end of a quint apparatus ladder when the extended aerial ladder pipe was pressurized.

One firefighter in command at a vocational school structure fire became ill while directing the deployment of firefighters and hoselines at the scene. His condition worsened after being taken to a local hospital and he died from a heart attack.

One firefighter was killed at the scene of a commercial occupancy structure fire when a brick facade at the front of the building became unstable and collapsed outward, pinning him as he was running away.

One firefighter fell into the fire-involved basement of a residential structure shortly after firefighters determined that the floor was softening on the first floor. They were attempting to evacuate the structure.

One firefighter died from smoke inhalation and thermal burns when he became caught and trapped inside a residential structure after sending another firefighter back to their apparatus to retrieve a tool and the fire inside of the structure intensified.

One firefighter was crushed by debris while working an incendiary fire in an abandoned house. While the firefighter was putting out hot spots in the attic, the roof collapsed.

One firefighter, while assisting with apparatus operations at the scene of a residential structure fire, was stricken with a heart attack as he attached a hoseline to a truck.

One firefighter died from smoke inhalation when a structural collapse occurred while he was working in the attic of a residential fire. Debris from the collapse struck the firefighter dislodging his helmet and facepiece.

One firefighter died from a heart attack while assisting with a firefight in a residential structure. He exited the structure and suddenly collapsed.

Table 8: Onscene Structure Fire by Activity 2004-2008 + Multiple Firefighter Fatalities

Cause of Fatal Injury	2004	2005	2006	2007	2008
Advance Hoseline	11	5	15	10	13
Other	0	2	0	0	1
Incident Command	2	0	1	2	2
Pump Operations	0	0	1	0	0
Search and Rescue	4	6	4	15	2
Scene Safety	1	1	0	1	0
Support	0	0	0	2	1
Setup	0	3	0	0	1
Ventilation	2	1	0	3	1
Water Supply	2	1	1	2	0
Total	22	19	22	35	21
Mulitple Fatality Incidents	3	2	2	5	2
Multiple Firefighter Fatality Deaths	6	4	4	17	4

Fireground Operations: Onscene Nonstructure Fire

Seven firefighters died while on the scene of seven different nonstructure fire incidents. Six of the seven deaths were related to wildland incidents including

- the crash of a SEAT;

- a fall while scouting a fire in extremely rough terrain and dangerous rock cliffs;

- trauma from being struck by a tree while on a felling team;

- smoke inhalation and thermal burns when overrun by flames; and

- two heart attacks, one while helping roll out a hoseline and the other while supervising a prison firefighting crew.

One firefighter was killed when a gunman apparently set a vehicle ablaze in order to draw responders into range so that he could shoot them from a barricaded single-family residence. Two law enforcement officers were also wounded in the incident.

Responding/Returning

Twenty-four firefighters died while responding to or returning from emergency incidents in 2008: 21 while responding to an emergency incident and 3 while returning from an emergency. Two of the incidents, both associated with wildland, were multiple firefighter fatalities: an air tanker crash taking the lives of three firefighters, and a bridge collapse killing two in a brush truck.

The types of nonaircraft fire department apparatus involved in vehicle collisions (one vehicle fall), killing eight firefighters, included three engines, two tankers (tender), one ambulance, and one brush truck (non-collision, fall due to structural collapse of a bridge undermined by fire). Five of the incidents were while responding and two while returning.

The status of seatbelt use in all six fire apparatus collisions was reported. Each of the six firefighters killed were operating the apparatus, three wore seatbelts. None of the three firefighters wearing seatbelts were ejected from the vehicle. All three firefighters without seatbelts were ejected from their vehicles.

Seatbelt use by the two firefighters killed in the brush truck fall/bridge collapse is unknown, but neither firefighter was ejected from the truck.

Speed was a factor in one of the six apparatus collisions; one occurred at an intersection; two incidents on a curved section of roadway; three on narrow roads; one at night and one at dawn; and one ambulance was struck head on by an oncoming vehicle (the operator killed was wearing a seatbelt).

Six firefighters were killed in crashes that involved personal vehicles, all while responding to an emergency. In at least four of the six fatalities, the firefighter was not wearing a seatbelt (two unknown). Three of the four firefighters without seatbelts were ejected from their vehicle.

Four of the six personal vehicles were operating with emergency lights (two unknown). Only one of the six personal vehicle crashes occurred at an intersection, two along curved sections of roadway, and three at night with one of the nighttime crashes also in poor weather conditions.

Excessive speed was cited as a factor in half (three) of the personal vehicle crashes while responding. Four firefighters died from heart attacks while responding to, and one while returning from an incident. One other firefighter died from a stroke while responding. None of these incidents involved a vehicle collision.

Table 9. Firefighter Deaths While Responding to or Returning From an Incident

Year	Number of Firefighter Deaths
2008	24
2007	26
2006	15
2005	22
2004	23
2003	36
2002	13
2001	23
2000	19
1999	26
1998	14

They Were Just Seventeen: Seatbelts and Speed

It is clear where mitigation/prevention efforts need to be focused in terms of younger-aged firefighter fatalities. The vast majority (92 percent) of young firefighters were riding in or operating motor vehicles (one bicycle) when they were killed. Of these firefighters, 85 percent were responding to an incident. This is in sharp contrast to just 14 percent of all other firefighters killed since 1990 while responding to an incident.

The youngest firefighter killed in 2008 turned 17 less than a month before being killed when responding from his residence in a privately-owned vehicle (POV) to an automatic fire alarm at approximately 0047 hours. The young man failed to negotiate a curve in the roadway striking an electrical utility box and several trees; the vehicle rolled several times. He was not wearing a seatbelt; was ejected from his vehicle; dying at the scene, and was discovered by utility workers several hours later.

In nearly 20 years, not including the firefighter fatalities associated with September 11, 2001, more than 6 out of every 1,000 onduty firefighter fatalities were age 17 or younger. When including the WTC firefighter deaths in the analysis, more than 5 out of every 1,000 firefighter deaths were 17 or younger.

Other Onduty Deaths

In 2008, 30 firefighters died on duty while engaged in activities that, with the exception of one incident that took the lives of 9 firefighters, were not on the scene of an emergency or associated with training. Eleven of the 30 were engaged in emergency duties.

- Twelve firefighters suffered heart attacks (9) or strokes (3) while on duty but not assigned to an incident or emergency response.

- One firefighter suffered a heart attack after responding to the station for an emergency call but stayed behind to staff the radios because he had begun to feel ill.

- Two firefighters died as the result of injuries that they received after a fall. One fell from a piece of heavy machinery while clearing fire breaks and the other from the loft of his fire station while fueling an emergency generator.

- One firefighter was fatally injured while operating an articulating aerial platform fire apparatus having just completed inspecting storm damages to the fire station with an insurance adjuster.

- Two firefighters died in separate incidents from medical emergencies: one from blood clots in the lungs and the other from a torn blood vessel in one leg.

- One firefighter died when his fire station accidently exploded.

- One firefighter died when the medical helicopter in which he was being transported to the hospital collided in midair with another medical helicopter.

- One fire department Chaplin died while performing his officially assigned duties.

- Nine firefighters died upon takeoff when their helicopter experienced a loss of power to the main rotor during takeoff, and subsequently impacted trees and terrain.

Training

In 2008, 12 firefighters died while engaged in training activities. Eight of the deaths were heart attacks:

- Two firefighters, one an instructor, experienced heart attacks while participating in two separate live fire training activities.

- Two firefighters suffered heart attacks during self-contained breathing apparatus (SCBA) drills at their respective fire academies.

- One firefighter's heart attack occurred in the station shortly after classroom training and another during what was reported to be particularly physically challenging training.

- One heart attack occurred while a firefighter operated a pump panel and was leading students through a flammable liquids fire scenario.

- One heart attack occurred while the firefighter was in transit to a training activity.

There were no heart attacks that occurred during or after fitness evaluations or physical fitness activities. While two firefighters died during physical fitness activities, one was from blood clots in the lungs and the second was from acidosis and dehydration.

One firefighter on training duty was shot and killed by a gunman who was randomly shooting at patrons and workers in a fast food establishment where he had stopped to pick up lunch.

One firefighter suffered a medical emergency while driving a tanker (tender) as a part of a training exercise.

Table 10: Firefighter Fatalities While Engaged in Training

Year	Number of Firefighter Deaths
2008	12
2007	11
2006	9
2005	14
2004	13
2003	12
2002	11
2001	14
2000	13
1999	3
1998	12

Nonfire Emergencies

After the Incident

Thirteen firefighters died after the conclusion of their onduty activity, 11 deaths were heart attacks, 1 a stroke, and 1 from trauma involving the accidental acceleration of a vehicle that crashed through a fire station bay door.

Type of Emergency Duty

In 2008, 75 firefighters died while responding to or working on the scene of an emergency. This number includes deaths resulting from injuries sustained on the incident scene or en route to the incident scene and firefighters who became ill on an incident scene and later died. It does not include firefighters who became ill or died while returning from an incident, e.g., a vehicle collision. Figure 8 shows the number of firefighters killed while responding to or working on the scene of an emergency in 2008.

Fifty-four firefighters were killed during firefighting duties; 4 firefighters were killed on emergency medical services (EMS) calls; 12 on motor vehicle accidents; 1 firefighter was killed in association with a technical rescue incident; and 4 were killed during other emergency circumstances.

Figure 8. Type of Emergency Duty

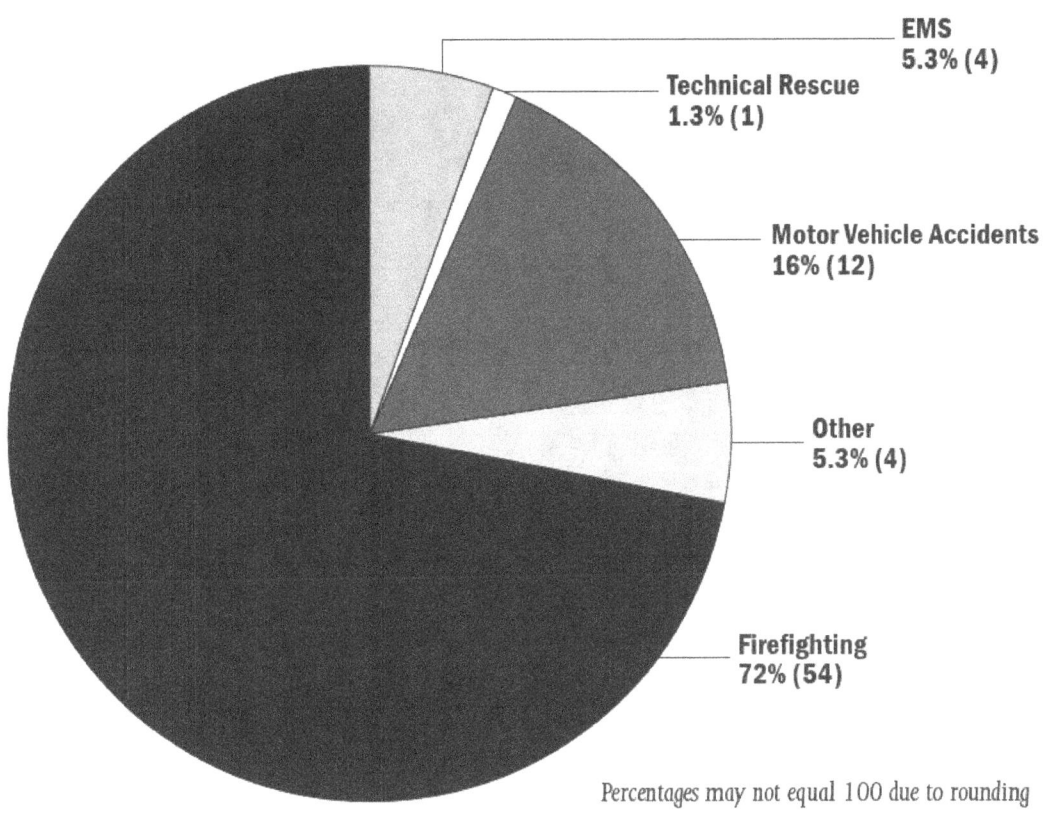

EMS
5.3% (4)

Technical Rescue
1.3% (1)

Motor Vehicle Accidents
16% (12)

Other
5.3% (4)

Firefighting
72% (54)

Percentages may not equal 100 due to rounding

Cause of Fatal Injury

The term "cause of injury" refers to the action, lack of action, or circumstances that resulted directly in the fatal injury. The term "nature of injury" refers to the medical cause of the fatal injury or illness which is often referred to as the physiological cause of death. A fatal injury usually is the result of a chain of events, the first of which is recorded as the cause.

Figure 9 shows the distribution of deaths by cause of fatal injury or illness.

Figure 9. Fatalities by Cause of Fatal Injury (2008)

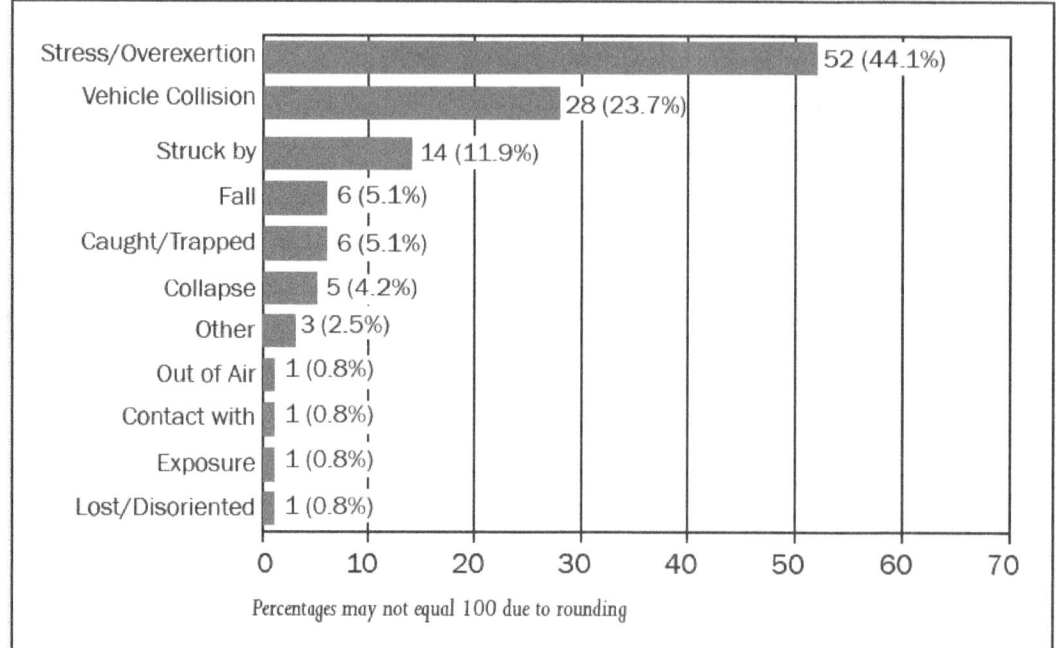

Percentages may not equal 100 due to rounding

Stress or Overexertion

Firefighting is extremely strenuous physical work and is likely one of the most physically demanding activities that the human body performs.

Stress or overexertion is a general category that includes all firefighter deaths that are cardiac or cerebrovascular in nature such as heart attacks, strokes, and other events such as extreme climatic thermal exposure. Classification of a firefighter fatality in this cause of fatal injury category does not necessarily indicate that a firefighter was in poor physical condition.

Fifty-two firefighters died in 2008 as a result of stress/overexertion:

- Forty-five of the stress deaths were heart attacks.

- Five firefighters died due to a CVA.

- One firefighter died from a tear in a blood vessel near his heart.

- One firefighter died from acidosis and dehydration.

22

Table 11: Deaths Caused by Stress or Overexertion

Year	Number	Percent of Fatalities
2008	52	44
2007	55	46.6
2006	54	50.9
2005	62	53.9
2004	66	56.4
2003	51	45.9
2002	38	38
2001	43	40.9*
2000	46	44.6
1999	56	49.5
1998	43	46.2

Does not include the firefighter deaths of September 11, 2001, in New York City.

Vehicle Crashes

After stress or overexertion, the perennial cause of fatal injury resulting in the most firefighter fatalities is vehicle crashes. Twenty-eight firefighters were killed in 2008 as a result of vehicle crashes. Fourteen of these deaths occurred in an aircraft crash, up from just one such fatality in 2007. Fourteen firefighters were killed in nonaircraft vehicle crashes.

- Eight of the nonaircraft crashes involved the firefighter's personal vehicle. One death occurred while performing scene safety at a MVA on an Interstate (struck by a semitractor-trailer); one motorcycle crash occurred while the firefighter was on fire department business; six deaths occurred while the firefighters were responding to an incident.

- Two crashes and two deaths involved a fire department tanker (tender).

- Four crashes and four deaths involved three engines and an ambulance.

- In all 14 of the nonaircraft vehicle crashes, the firefighter killed was operating the vehicle. No seatbelt was used in 8 of the 11 cases where seatbelts were available and the status of their use was known. Of the eight drivers not wearing seatbelts, six were fully ejected from their vehicles. In two crashes, the status of seatbelt use is unknown or not reported (nonapplicable for motorcycles).

Figure 10. Firefighter Fatalities in Vehicle Collisions

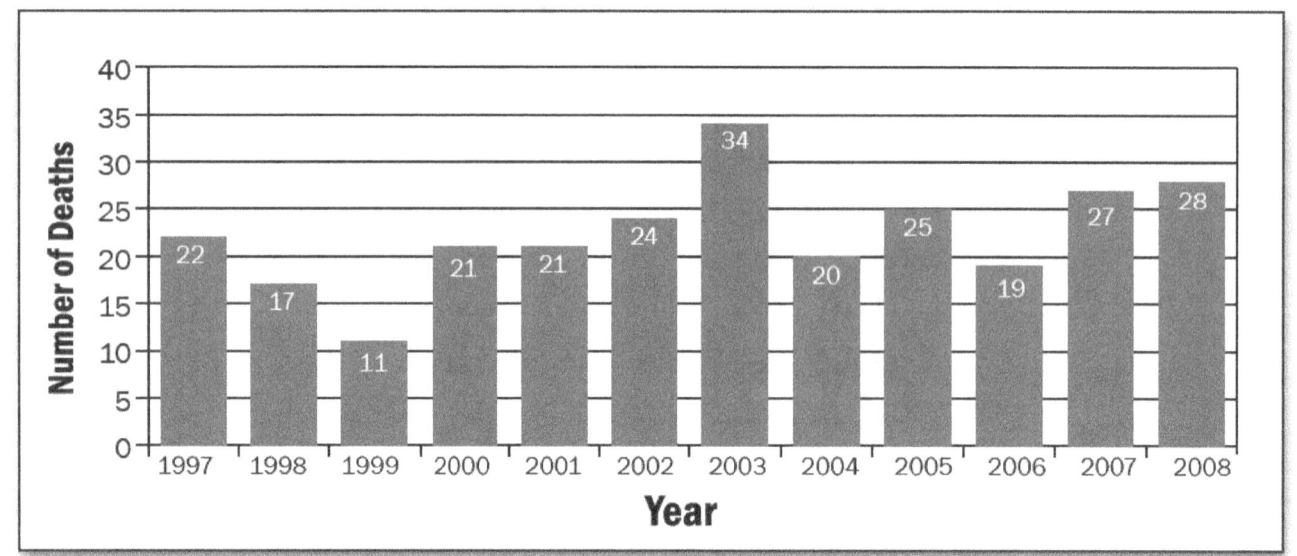

Lost or Disoriented

One firefighter died in 2008 when he became lost or disoriented inside a residential structure fire. The firefighter was on the first engine to arrive on the scene. He and another firefighter advanced a hoseline to the front door of the residence. One of the firefighters was sent back to the fire truck for a tool. When he returned, his partner was gone and the nozzle remained by the doorway. At the same time, the fire inside the structure intensified.

Caught or Trapped

Six firefighters were killed in five incidents when they were caught or trapped in 2008. This classification covers firefighters trapped in wildland and structural fires who were unable to escape due to rapid fire progression and the byproducts of smoke, heat, toxic gases, and flame. This classification also includes firefighters who drowned, and those who were trapped and crushed.

- One firefighter was trapped while conducting a search of a duplex residence when the water supply to the fire scene was interrupted due to frozen hydrants.

- One firefighter on the nozzle of a line that was advanced into the basement of a residence was trapped and overcome by heat conditions when trying to escape and encountered a blocked door.

- Two firefighters of a four person crew were caught and trapped as fire conditions rapidly changed due to the collapse of an interior wall of a large 79,000 square foot manufacturing occupancy structure.

- One firefighter was caught by rapid fire progress after he and another firefighter, both on the first truck to arrive on scene, advanced a charged 1-3/4 inch preconnected handline into a residential structure, but then had difficulty locating the fire as smoke and heat conditions intensified. The two firefighters became separated and rapid fire progress occurred.

- One wildland firefighter was caught by a rapidly advancing flame front while scouting an area under his supervision.

Collapse

Five firefighters in four incidents died in 2008 as the result of structural collapses.

- Two firefighters in the basement of a residential structure were buried under collapsed structural components when a large portion of the first floor collapsed.

- One firefighter was operating a nozzle at a doorway of a commercial occupancy structure when a two-story brick façade collapsed outward, pinning him as he was running away.

- One firefighter was crushed by debris while putting out hotspots in the attic of an abandoned residential structure when the roof collapsed.

- One firefighter's helmet and facepiece were dislodged by a collapse while working in the attic of a residential structure fire.

Struck by Object

Being struck by an object was the third leading cause of fatal firefighter injuries in 2008. Fourteen firefighters died in 2008 as the result of being struck by an object, almost three times the number in 2007 during which five firefighters died from this cause.

- Five firefighters were struck by vehicles in 2008. Four of these were at the scene of vehicle crashes and one occurred when a POV accidently accelerated through the bay door of a station.

- There were three separate incidents in 2008 where three firefighters were killed when they were struck by bullets.

- Two firefighters died in structural explosions. One of these incidents involved a room that housed electrical equipment, while the other involved a firehouse filled with propane gas.

- Two firefighters died after being struck by aerial apparatus. During one incident, a motorized water monitor and 30 feet of aluminum pipe were projected out the end of an extended pressurized aerial ladder and struck a firefighter. During the other incident, a firefighter was crushed between the aerial device and apparatus as he bedded the aerial device.

- One firefighter was killed when his position on a storm-watch assignment was struck by a tornado and one wildland firefighter was struck by a tree.

SHOOTINGS: With three incidents and deaths, the year 2008 was the deadliest year for firefighter shootings since 2000, when three firefighters were killed in two incidents. Only in 1996 did more firefighters die from gunshot wounds in a single year when four firefighters were killed in one incident, and another firefighter was killed in a separate incident. In 2002, 2004, and 2005, one firefighter was lost each year as a result of being shot.

Fall

Six firefighters died in 2008 as the result of falls.

- One firefighter fell from the loft of a fire station while fueling a generator.

- Two firefighters fell while responding to a wildfire in a brush truck when the bridge they were crossing, undermined by fire, suddenly collapsed.

- One firefighter fell into the fire-involved basement of a residential structure and died from positional asphyxiation when he was crushed by debris, principally a couch, and was unable to breathe.

- One firefighter fell from a dangerous cliff while scouting a wildland fire in extremely rough terrain.

- One firefighter died when he struck his head after falling off of a grader he was repairing during wildland operations to improve road conditions and access for firefighters.

Out of Air

One firefighter died of smoke inhalation and thermal burns on the 14th floor of a highrise residential occupancy. The firefighter ran out of air less than 20 minutes after donning his facepiece and was unable to exit the apartment before being overcome.

Contact With

One firefighter was electrocuted while in the platform of a tower ladder as he ascended through powerlines to position the ladder in a spot that would allow access to the upper floors of a residence. A utility company worker had arrived at the scene and departed before the fatal incident took place.

Exposure

One firefighter died from a heart attack with cause of death listed as hyperthermia and dehydration; a reported intestinal condition may have contributed to his illness.

Other

Three firefighters died in 2008 of a cause that is not categorized above.

- Two firefighters died in separate incidents from medical emergencies: one from blood clots in the lungs and the other from a torn blood vessel in one leg.

- One firefighter died when his preexisting medical condition caused his tanker to be involved in a crash and to rollover.

Nature of Fatal Injury

Figure 11 shows the distribution of the 118 firefighter deaths that occurred in 2008 by the medical nature of the fatal injury or illness. For heart attacks, Figure 12 shows the type of duty involved.

Figure 11. Fatalities by Nature of Fatal Injury (2008)

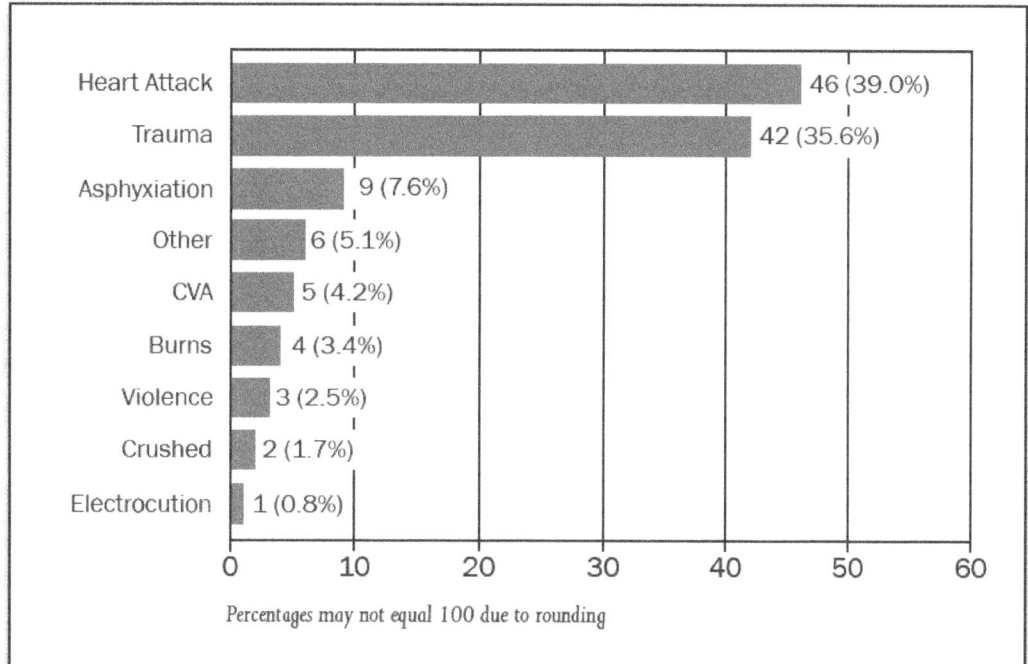

Percentages may not equal 100 due to rounding

Figure 12. Heart Attacks/Type Of Duty 2008

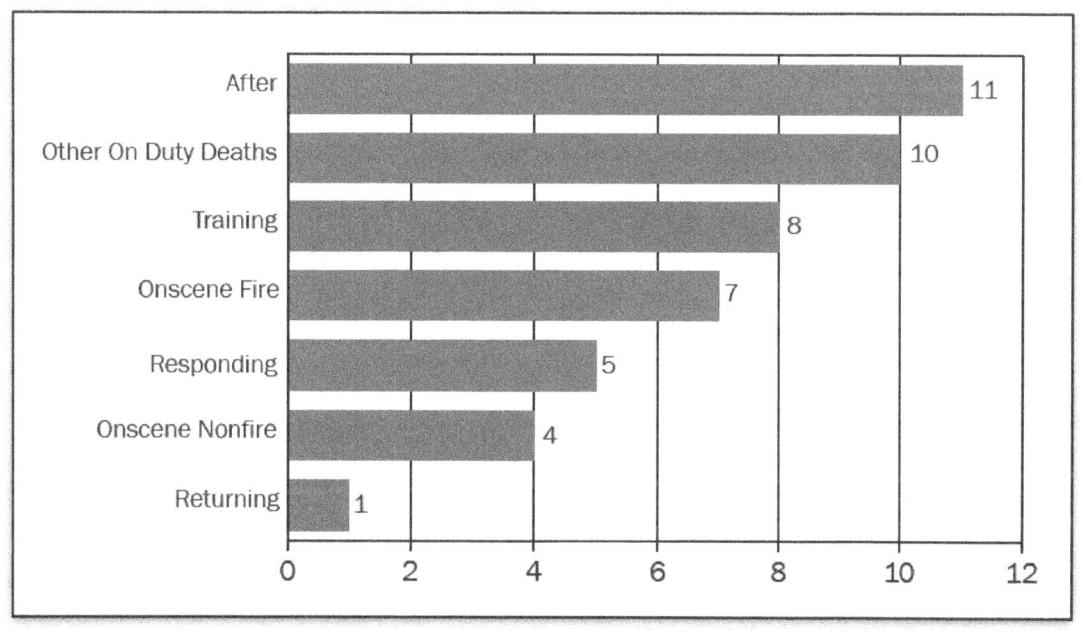

FIREFIGHTER AGES

Figure 13 shows the percentage distribution of firefighter deaths by age and nature of the fatal injury. Table 12 provides a count of firefighter fatalities by age and the nature of the fatal injury.

Younger firefighters were more likely to have died as a result of traumatic injuries such as injuries from an apparatus accident or becoming caught or trapped during firefighting operations. Stress-related deaths are rare below the 31 to 35 years of age category and they often include underlying medical conditions. The youngest firefighter to die of a cardiac-related cause in 2008 was age 32. A 28 year-old firefighter died from a heart attack with cause of death listed as hyperthermia and dehydration; a reported intestinal condition may have contributed to his illness. Statistics show that stress plays an increasing role in firefighter deaths as age increases.

Figure 13. Fatalities by Age and Nature (2008)

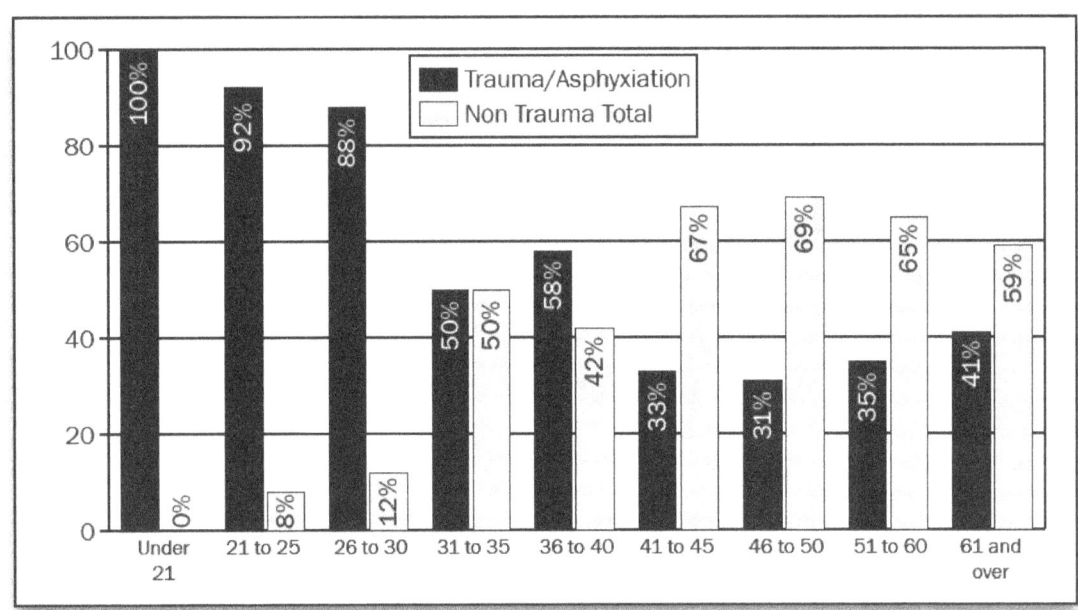

Table 12: Firefighter Ages and Nature of Fatal Injury (2008)

Age Range	Non Trauma Total	Trauma/Asphyxiation Total
under 21	0	6
21 to 25	1	12
26 to 30	1	7
31 to 35	6	6
36 to 40	5	7
41 to 45	10	5
46 to 50	9	4
51 to 60	11	6
61 and over	13	9

The youngest firefighter killed on duty in 2008 was Firefighter Roy Dale Smith, III, of Virginia. He was responding from his residence in a POV to an automatic fire alarm when he was involved in a crash. He was 17 years old.

The oldest firefighter killed on duty in 2008 was Fire Police Officer Honorary Chief Edward A. Junginger of New York. He died of a heart attack while helping to reopen a local highway after an automobile crash. He was 82 years old.

FIXED PROPERTY USE FOR STRUCTURAL FIREFIGHTING DEATHS

There were 21 fatalities in 2008 where firefighters became ill or injured while on the scene of a structure fire. Table 13 shows the distribution of these deaths by fixed property use. In most years, residential occupancies accounted for the highest number of these fireground fatalities.

Table 13: Structural Firefighting Deaths by Fixed Property Use in 2008

Residential	15	71.4
Commercial	4	19.0
Manufacturing	1	4.8
Educational	1	4.8

TYPE OF FIREGROUND ACTIVITY

In 2008, there were a total of 28 firefighter deaths on the fireground. Table 14 shows the types of fireground activities in which firefighters were engaged at the time they sustained their fatal injuries or illnesses. This total includes all firefighting duties, such as wildland firefighting and structural firefighting.

Table 14: Type of Activity (2008)

Activity	Number
Fire Attack	16
Search and Rescue	2
Setup and Support	3
Ventilation	1
Command	4
Other	2
Total	28

Table 15: Firefighter Deaths While Engaged in Fire Attack (2008)

Year	Number of Firefighter Deaths
2008	16
2007	19
2006	24
2005	11
2004	16
2003	11
2002	13
2001	13
2000	13
1999	16
1998	18

DEATHS BY TIME OF INJURY

The distribution of all 2008 firefighter deaths according to the time of day when the fatal injury occurred is illustrated in Figure 14. The time of fatal injury for seven firefighters was either unknown or unreported.

Figure 14. Fatalities by Time of Fatal Injury (2008)

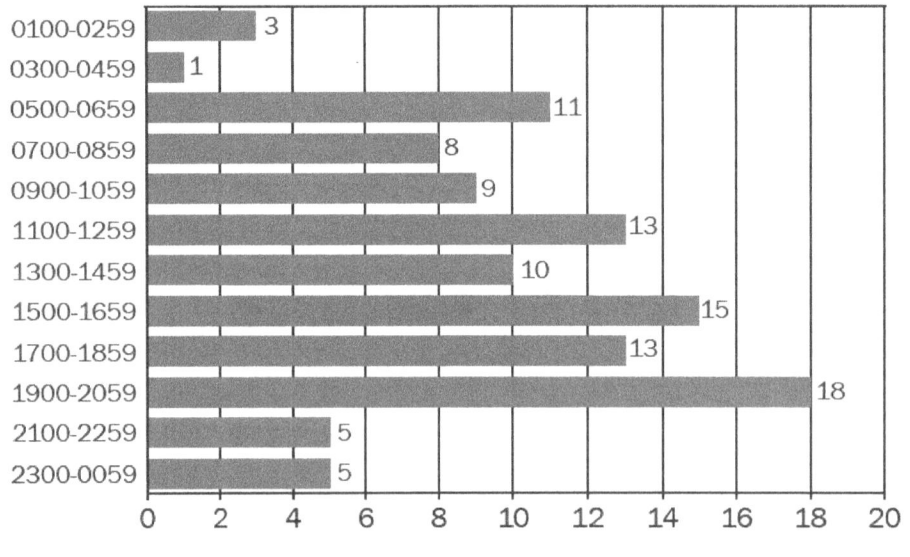

Figure 15 illustrates the 2008 firefighter fatalities by month of the year.

DEATHS BY MONTH OF YEAR

Figure 15. Deaths by Month of Year (2008)

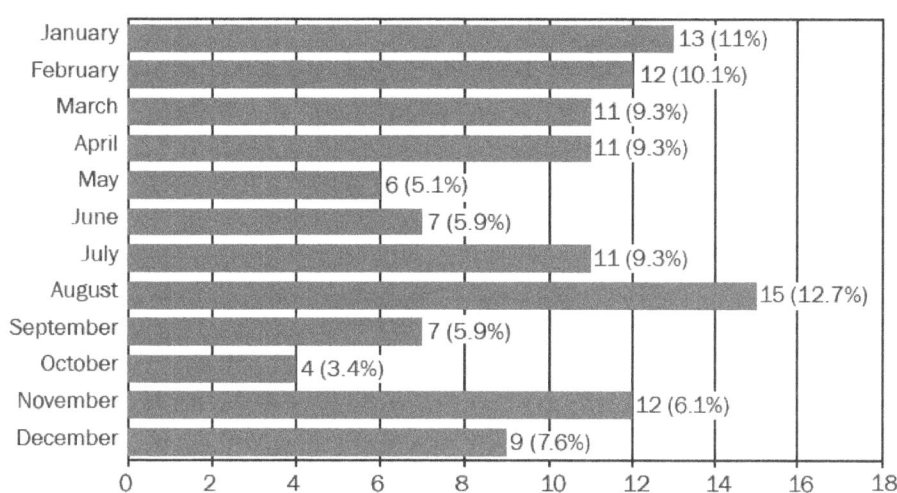

STATE AND REGION

The distribution of firefighter deaths in 2008 by State is shown in Table 16. Firefighters based in 36 States died in 2008.

The highest number of firefighter deaths, based on the location of the fire service organization in 2008, occurred in North Carolina with 11 deaths. Pennsylvania and Oregon had the next highest total of firefighter fatalities in 2008 with 9 deaths each. In 2007, South Carolina experienced the highest loss of firefighters with 11 deaths.

31

Table 16: Firefighter Fatalities by State by Location of Fire Service* (2008)

State	Fatalities	Percentage
Alabama	2	1.7
Arkansas	1	0.8
Arizona	2	1.7
California	7	5.9
Colorado	4	3.4
Connecticut	1	0.8
Delaware	1	0.8
Florida	2	1.7
Georgia	2	1.7
Hawaii	1	0.8
Illinois	6	5.1
Indiana	2	1.7
Kentucky	1	0.8
Louisiana	3	2.5
Maryland	3	2.5
Maine	1	0.8
Michigan	2	1.7
Minnesota	1	0.8
Missouri	7	5.4
Mississippi	3	2.5
Montana	5	4.2
North Carolina	11	9.3
New Jersey	2	1.7
New Mexico	1	0.8
New York	7	5.9
Ohio	6	5.1
Oklahoma	1	0.8
Oregon	9	7.6
Pennsylvania	9	7.6
Rhode Island	2	1.7
South Carolina	2	1.7
Texas	4	3.4
Virginia	2	1.7
Washington	2	1.7
Wisconsin	2	1.7
West Virginia	1	0.8

* This list attributes the deaths according to the State in which the fire department or unit is based, as opposed to the State in which the death occurred. They are listed by those States for statistical purposes and for the National Fallen Firefighters Memorial at the NETC.

Figure 16. Firefighter Fatalities by Region (2008)

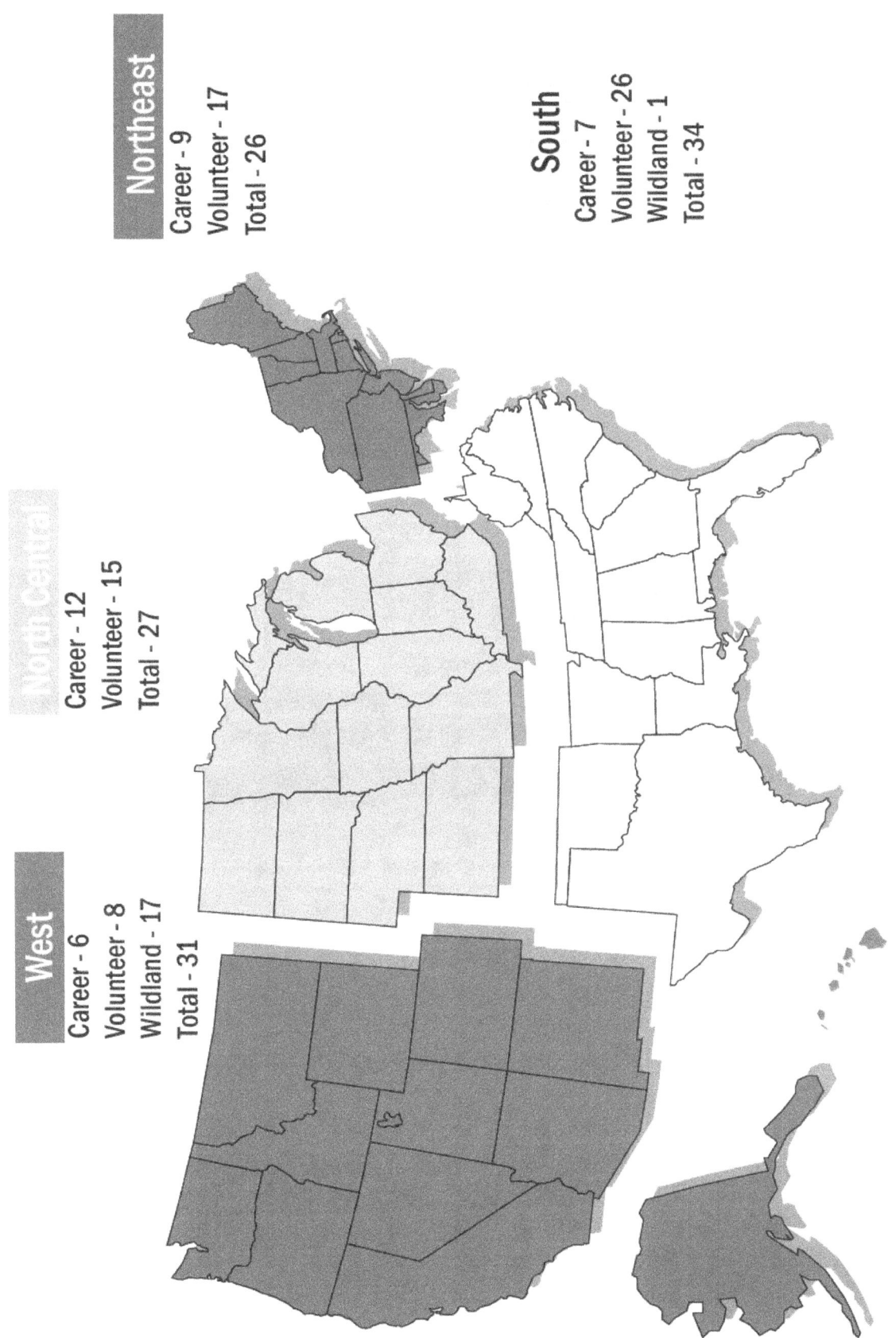

Northeast
Career - 9
Volunteer - 17
Total - 26

South
Career - 7
Volunteer - 26
Wildland - 1
Total - 34

North Central
Career - 12
Volunteer - 15
Total - 27

West
Career - 6
Volunteer - 8
Wildland - 17
Total - 31

Figure 17. Onduty Firefighter Fatalities 2008 by Fire Department Location

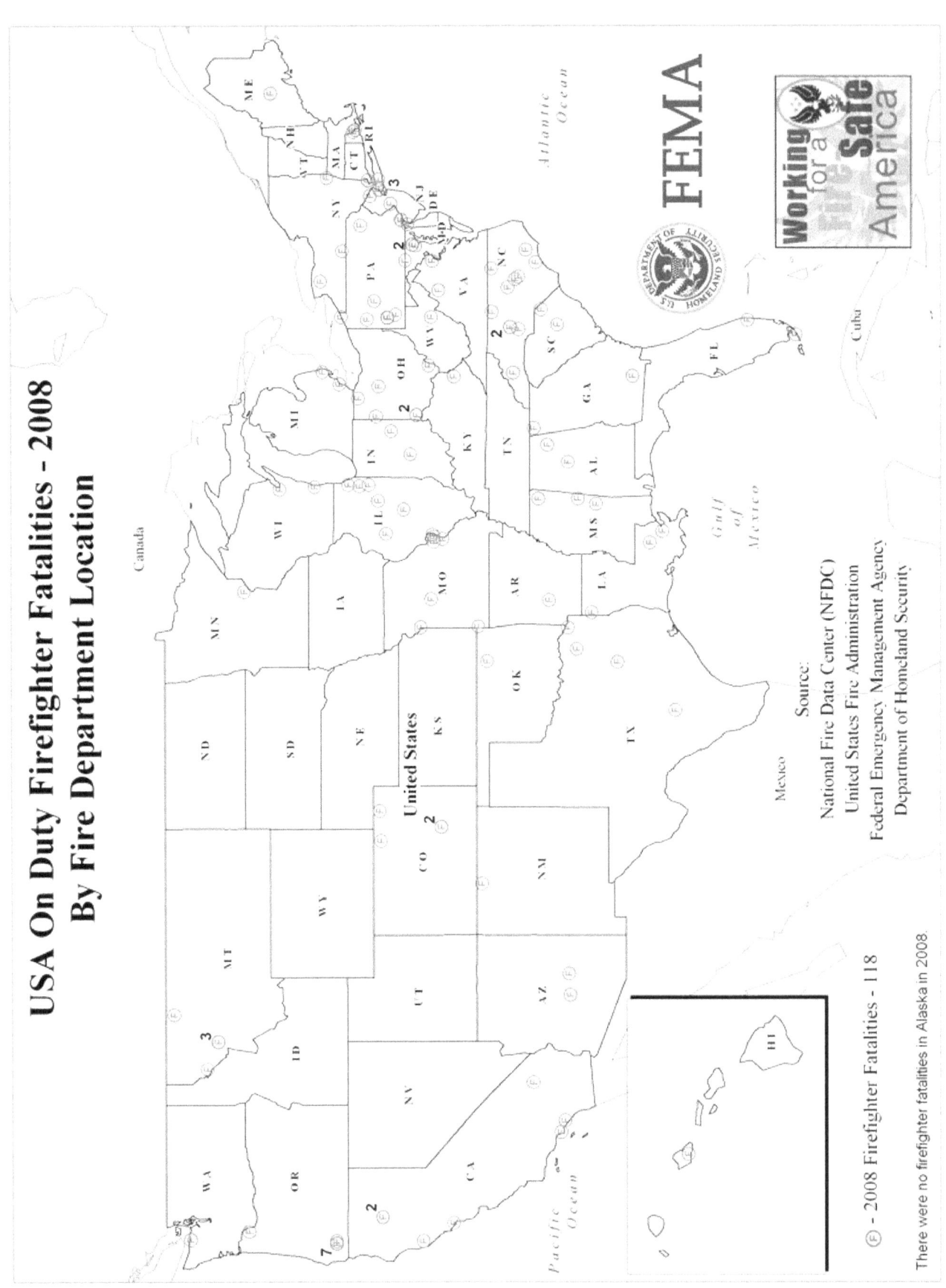

Figure 18. Onduty Firefighter Fatalities 2008 by Incident Location

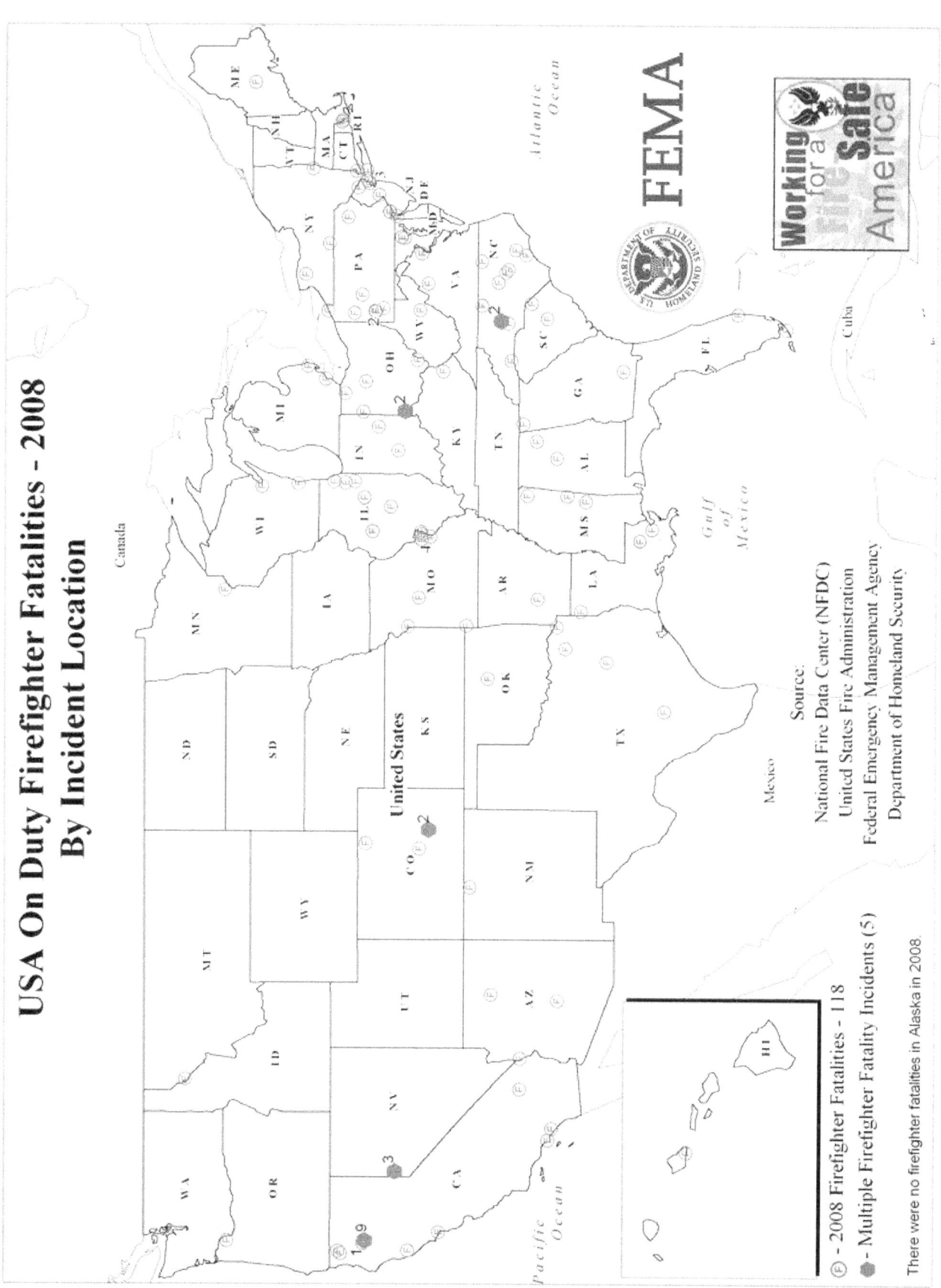

ANALYSIS OF URBAN/RURAL/SUBURBAN PATTERNS IN FIREFIGHTER FATALITIES

The United States Census Bureau defines "urban" as a place having a population of at least 2,500 or lying within a designated urban area. "Rural" is defined as any community that is not urban. "Suburban" is not a census term but may be taken to refer to any place, urban or rural, that lies within a metropolitan area defined by the Census Bureau, but not within one of the central cities of that metropolitan area.

Fire department areas of responsibility do not always conform to the boundaries used by the Census Bureau. For example, fire departments organized by counties or special fire protection districts may have both urban and rural coverage areas. In such cases, where it may not be possible to characterize the entire coverage area of the fire department as rural or urban, firefighter deaths were listed as urban or rural based on the particular community or location in which the fatality occurred.

The following patterns were found for 2008 firefighter fatalities. These statistics are based on answers from the fire departments and, when no data from the departments were available, the data are based upon population and area served as reported by the fire departments.

Table 17: Firefighter Deaths by Coverage Area Type (2008)

	Urban/Suburban	Rural	Federal or State Parks/Wildland	Total
Firefighter Deaths	65	35	18	118

Appendix A

In memory of all firefighters

who answered their last call in 2008

To their families and friends

To their service and sacrifice

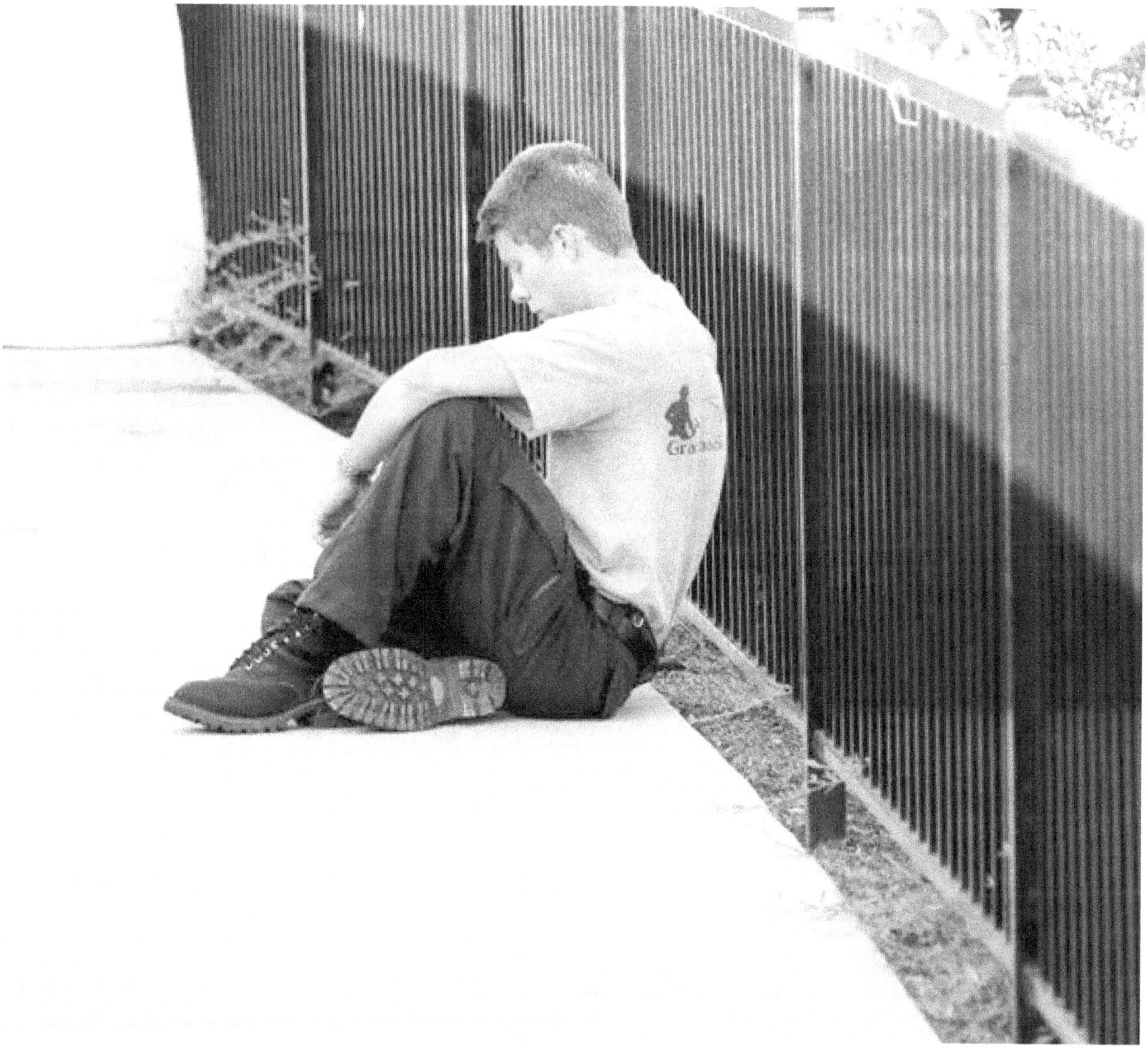

January 1, 2008–0616 hrs
Donald Gerald Paterson, Firefighter
Age 65, Volunteer
Kimball Township Fire Department, Michigan

After a night of heavy snowfall, Firefighter Paterson and the members of his fire department were dispatched to a report of wires down. Firefighter Paterson arrived at the fire station and assumed the role of dispatcher at the station's radio console.

While other firefighters attended to the incident, Firefighter Paterson monitored the radio and used a snowblower to remove accumulated snow from the path that led to the fire station. Firefighter Paterson commented to another firefighter that he was having slight shoulder pain and chest discomfort.

At the conclusion of the incident, Firefighter Paterson returned home. At approximately 1400 hrs, Firefighter Paterson became ill. An ambulance transported him to a local hospital where he was diagnosed as suffering from a heart attack. Firefighter Paterson was rushed into surgery at the hospital but did not recover.

January 1, 2008–0730 hrs
Paul Lewis Ellington, Firefighter
Age 36, Volunteer
Oregon Hill Volunteer Fire Department, North Carolina

Firefighter Ellington was responding to a structure fire in his personal vehicle, a pickup truck.

The vehicle hit a stop sign, left the roadway, rolled over, struck a group of trees and went down an embankment. Firefighter Ellington, who was not wearing his seatbelt at the time of the crash, was ejected and trapped under the vehicle when it came to rest.

Law enforcement officials investigating the crash said that speed played a role in the incident. Firefighter Ellington was pronounced dead at the scene.

January 3, 2008–1954 hrs
John H. Martinson, Lieutenant
Age 40, Career
Fire Department City of New York, New York

Lieutenant Martinson and the members of his engine company were dispatched as part of a full assignment to a report of a fire in a 25-story highrise residential occupancy structure.

Upon their arrival on the scene, Lieutenant Martin and his crew used the firefighter's feature to ride an elevator to the floor below the reported fire. He and his crew used the stairs to ascend to the 14th floor. When they opened the door from the stairwell to the fire floor, firefighters discovered heavy smoke.

Continued on next page

38

Lieutenant Martin donned his SCBA and entered the hallway to look for the location of the apartment that was on fire.

The floor layout for the building was confusing and the apartment numbering system was not clear which made finding the apartment that was experiencing the fire difficult. Once the apartment was discovered, firefighters advanced a hoseline and began to apply water to the fire. The fire was intensified by a strong wind.

Firefighters working in the fire apartment discovered Lieutenant Martinson on the floor of the apartment, approximately 3 feet from the doorway. He was unconscious and his facepiece and helmet had been removed. Firefighters rushed Lieutenant Martinson to the stairwell and to a lower floor.

Emergency medical procedures were initiated and Lieutenant Martinson was transported to the hospital. He was pronounced dead at the hospital due to smoke inhalation and thermal burns.

A report prepared by the fire department concluded that Lieutenant Martinson ran out of air less than 20 minutes after donning his facepiece and was unable to exit the apartment before being overcome. The indirect causes of his death were cited as fireplay by a child, the failure of the apartment occupants to close the apartment door when they evacuated, failure to team up with another member while operating in an Immediately Dangerous to Life and Health (IDLH) atmosphere, and failure to leave the IDLH atmosphere when the SCBA low air indicators activated.

January 6, 2008–0742 hrs
James L. Robeson, Captain
Age 50, Career
Scranton Fire Department, Pennsylvania

At 0714 hours, Captain Robeson and the members of his fire department were dispatched to a fire in a residence. Upon arrival, firefighters found a working fire. The Incident Commander (IC) requested the response of the local electrical utility company to the scene. A utility company worker arrived at the scene and then departed.

Captain Robeson and another firefighter were in the platform of a tower ladder, ascending through powerlines to position the ladder in a spot that would allow access to the upper floors of the residence or for ladder pipe operations. At approximately 0742 hours, Captain Robeson and three other firefighters were electrocuted. Power from a 13,000 volt electrical line either arced to the ladder or the ladder came in contact with one of the lines.

After power was shut off, Captain Robeson and the other firefighter in the platform were removed and treated. Captain Robeson could not be revived.

Two residents of the home also died as a result of the fire. It is unknown if the electrical utility worker was asked to cut power to the lines in the street or just from the street to the house. The incident is currently under investigation.

January 7, 2008–0836 hrs
Harvey Jordan, Firefighter
Age 71, Volunteer

Penn Hills #1 Volunteer Fire Department Co. No. 1, Pennsylvania

Firefighter Jordon responded to the fire station in his personal vehicle to a report of a structure fire. When he arrived at the fire station, he told other firefighters that he was short of breath. Firefighters, including a paramedic, began to assess his condition and provide treatment.

Firefighter Jordon became unconscious. Firefighters applied an Automated External Defibrillator (AED) and began Cardiopulmonary Resuscitation (CPR). Firefighter Jordon was transported to a local hospital but did not survive. The cause of death was listed as a heart attack.

January 9, 2008–1415 hrs
Jarrett Aliber Dixon, Fire Apparatus Driver Operator
Age 36, Career

Baltimore County Fire Department, Maryland

Fire Apparatus Driver Operator Dixon was exercising on a treadmill while on duty at his fire station on January 9, 2008. He began to experience chest pains and went to a fire department ambulance based in his station. He hooked himself up to a cardiac monitor and told other firefighters that he needed to go to the hospital. Fire Apparatus Driver Operator Dixon was transported to the hospital.

Fire Apparatus Driver Operator Dixon was treated at the hospital and referred for further treatment. On January 12, 2008, at approximately noon, Fire Apparatus Driver Operator Dixon became ill at home. An ambulance transported him back to the hospital where he died later that day. The cause of death was listed as blood clots in the lungs.

January 16, 2008–0950 hrs
Johnny Bajusz, Firefighter
Age 69, Volunteer

Layton Volunteer Fire Department—Monroe County Fire/Rescue, Florida

Firefighter Bajusz was responding to a recreational vehicle fire connected to a mobile home on Conch Key in his POV. During the response, he attempted a U-turn and was struck by another vehicle coming from behind.

Firefighter Bajusz was trapped in his vehicle and had to be extricated by firefighters. He was flown by medical helicopter to a regional trauma hospital.

Firefighter Bajusz suffered serious injuries with ongoing complications and remained in the hospital for just over 6 months. He asked to be taken to his home where he passed away the following day. Firefighter Bajusz died on July 18, 2008 as the result of an infection.

January 17, 2008–Time Unknown
Louis Paul "Lou" Berra, Firefighter/Paramedic
Age 49, Career
West County EMS and Fire Protection District, Missouri

Firefighter/Paramedic Berra reported for duty on January 17, 2008. During the shift, he participated in station and apparatus maintenance duties and attended classroom training on ice rescue operations. Firefighter/Paramedic Berra also participated in department-mandated physical fitness training during his shift, running on a treadmill for over 45 minutes.

At approximately 2130 hours, Firefighter/Paramedic Berra retired to his dorm room in the fire station. The next morning, other firefighters found him deceased in his dorm room. The cause of death was listed as heart disease.

January 18, 2008–1654 hrs
Herman Sylvester Jones, Lieutenant
Age 58, Career
Raleigh Fire Department, North Carolina

Lieutenant Jones was on duty in his fire station and acting as the Company Officer (CO) for Engine 10. His company responded to an emergency medical incident at 1047 hours. At 1300 hours, the company Captain arrived from a class and Lieutenant Jones assumed the duties of driver operator.

At 1616 hours, Engine 10 responded to another emergency medical incident and then returned to quarters. Lieutenant Jones took a portable radio and went to the park behind the fire station to exercise. At 1654 hours, Engine 10 was dispatched to a report of a shooting. Lieutenant Jones sprinted approximately 100 yards from the park in order to respond. When he arrived at the apparatus, he was winded and complained of pain in his leg.

At the conclusion of the incident, Engine 10 attended to some nonemergency duties and then returned to quarters. At 1123 hours, Lieutenant Jones called the fire station phone from the fire station's day room. He told the CO that he was in pain and needed assistance. The Captain found Lieutenant Jones in distress complaining of severe leg pain.

An ambulance was dispatched and Lieutenant Jones was transported to the hospital. He was found to have a tear in a blood vessel near his heart and surgery was performed. His condition worsened, and Lieutenant Jones died on January 22, 2008.

January 20, 2008–1008 hours
Robert L. McAtee, Sr., Firefighter
Age 55, Volunteer
Huttonsville-Mill Creek Volunteer Fire Department, West Virginia

Firefighter McAtee responded to the fire station and drove an apparatus to the scene of a chimney fire. Upon arrival, he reported to the Fire Chief, his son, and told him that he thought that he pulled a muscle opening the apparatus bay door when leaving the station. He was later observed holding his chest and, when asked, indicated that he thought that perhaps it was the extremely cold weather that was causing him discomfort. He declined to be sent for evaluation or treatment and returned the apparatus to the station.

He returned home and was witnessed by his wife again holding his chest. At 1629 hours, his son was summoned to Firefighter McAtee's residence and found him unresponsive. CPR was initiated on scene by the Fire Chief and care was continued through transport to the local hospital, where Firefighter McAtee succumbed to his injuries.

January 22, 2008–0900 hrs
Christa Dawn Burchett, Assistant Fire Chief/EMS Director
Age 33, Career
Paintsville Fire-Rescue-EMS, Kentucky

Assistant Chief Burchett and another firefighter responded to a motor vehicle crash on a local highway. Upon arrival, they found that a passenger vehicle had crashed after sliding on ice on the roadway. The driver of the car, who was pregnant, appeared to be uninjured but asked to be transported to the hospital to be checked.

As the driver of the car was loaded into the ambulance, a tractor-trailer truck loaded with coal slid on the ice and struck the rear of the ambulance. Assistant Chief Burchett was struck by the truck and thrown under the ambulance. She was extricated from the crash and transported to the hospital, but was pronounced dead at 1004 hours.

The driver of the car involved in the original crash was also killed.

January 26, 2008–0125 hours
Walter Clyde Walker, Jr., Fire Chief
Age 68, Volunteer
Collinsville Volunteer Fire Department, Mississippi

Chief Walker was responding in his personal vehicle to a report of a vehicle rollover crash. As he responded, Chief Walker's vehicle left the right side of the roadway, crashed through a road sign, traversed a culvert, and struck a large tree. Chief Walker was not wearing a seatbelt and was killed in the crash due to massive chest and head trauma.

January 31, 2008–1400 hrs
Paul William "Rosie" Swander, Fire Chief
Age 74, Volunteer
Ohio City Fire Department, Ohio

Chief Swander went to his fire station at approximately 1400 hours to make preparations for the forecasted arrival of a winter storm. He was in the loft of the fire station fueling a generator when he fell from the loft to the floor below.

When Chief Swander failed to arrive at home for dinner, his wife called their son, a local sheriff's department employee, to let him know that Chief Swander was missing. Chief Swander's son found him in the fire station at approximately 1800 hours.

Chief Swander was flown by medical helicopter from a local hospital to a regional trauma center. He died on February 2, 2008, as a result of head injuries.

February 1, 2008–1000 hrs
Matthew P. Hubly, Fire Chief
Age 43, Paid-on-Call
Kankakee Township Fire Protection District, Illinois

Chief Hubly responded with the members of his fire department to a motor vehicle crash on the evening of January 31, 2008. The department's response was cancelled by another fire department.

The morning of February 1, 2008, Chief Hubly assisted another firefighter with clearing snow from areas around the fire station that could not be reached by a snow plow.

At 1000 hours, Chief Hubly and his assistant chief went to a meeting at a local industrial facility. The purpose of the meeting was to discuss fire protection issues at the facility. Chief Hubly became ill during the meeting and agreed to go to the hospital with his assistant chief.

Chief Hubly became unconscious in the car. Fire department and emergency medical units were dispatched. Chief Hubly was treated and transported to the hospital.

Chief Hubly died on February 7, 2008, as the result of a heart attack.

February 1, 2008–1430 hrs
Donald W. Hubbel, Captain
Age 42, Career

Baltimore City Fire Department, Maryland

Captain Hubbel worked his regular shift on January 31, 2008. The shift included strenuous rescue company training operations at the fire academy during the day.

On the morning of February 1, 2008, Captain Hubbel went off duty. At approximately 1430 hours, he was exercising on a treadmill at home and suffered a fatal heart attack.

February 4, 2008–1550 hrs

David Thomas Sherfick, Captain
Age 40, Career

Brown Township Fire-Rescue, Indiana

Captain Sherfick was the driver of an ambulance. The unit was returning from a patient transport to Indianapolis. The lights and siren on the ambulance were not in use.

As he drove on a local roadway, Captain Sherfick observed an oncoming vehicle drifting over the center line into his lane. He steered the ambulance to the right to attempt to avoid a collision but the other vehicle continued to drift and the vehicles crashed head-on.

Captain Sherfick was trapped in the ambulance and had to be extricated. He died as the result of traumatic injuries. Captain Sherfick and the ambulance's front seat passenger were wearing seatbelts. The passenger received minor injuries. The driver of the other vehicle also died and may have suffered some sort of medical emergency prior to the crash.

February 11, 2008–1924 hrs
James Earl Arthur, Firefighter
Age 19, Volunteer

Cold Water Fire & Rescue, North Carolina

Firefighter Arthur was responding in his personal vehicle, a half-ton pickup, to the report of a vehicle crash with a confirmed need for extrication.

As his vehicle passed through a curve in the road at high speed, Firefighter Arthur overcorrected to the left, went off of the road momentarily, then overcorrected to the right. The vehicle left the right side of the roadway and rolled several times. He was ejected from the vehicle during the crash. Firefighter Arthur was pronounced dead at the hospital as the result of traumatic injuries.

Firefighter Arthur was not wearing a seatbelt at the time of the crash.

February 12, 2008–1500 hrs
Kerry R. Sheridan, Fire Chief
Age 75, Career
Troy Township Fire Protection District, Illinois

On February 11, 2008, at 1513 hours, Chief Sheridan responded to the fire station from his home in the emergency mode due to multiple vehicle crashes that had occurred in his district. Upon his arrival at the fire station, Chief Sheridan directed the response of multiple fire department units to these emergencies.

On February 12, 2008, Chief Sheridan was in the fire station and directed units for responses to a vehicle crash and a mutual-aid fire incident. Later that afternoon, Chief Sheridan was discovered by staff in his office in full cardiac arrest. He was treated by onduty paramedics and transported to the hospital but did not survive.

February 16, 2008–1254 hrs
Vance Roland Tomaselli, Captain
Age 60, Paid-on-Call
San Bernardino County Fire Department, California

Captain Tomaselli was on duty in his fire station. Units from his station were dispatched to a fire investigation. When the incident nature was changed to a structure fire, Captain Tomaselli responded to the incident in a brush engine.

During the response, Captain Tomaselli suffered a stroke. The engine left the roadway and sideswiped a tree. He was able to continue his response to the scene and was treated by firefighters on the scene of the structure fire upon his arrival. He was transported by medical helicopter to the hospital. Captain Tomaselli died as a result of the stroke on February 21, 2008.

February 17, 2008–1955 hrs
Larry J. Lockhart, Firefighter
Age 69, Volunteer
Dayton District Volunteer Fire Department, Pennsylvania

Firefighter Lockhart and the members of his fire department responded to their fire station as the result of a reported structure fire. When he arrived at the fire station, Firefighter Lockhart told other firefighters that he was not feeling well and decided not to drive the apparatus. He went into the station's radio room.

Firefighter Lockhart collapsed inside the radio room. He was treated and transported to the hospital but was later pronounced dead of a heart attack.

February 19, 2008–1213 hrs
Michael James Hays, Firefighter
Age 64, Volunteer

Brazos Canyon Volunteer Fire Department, New Mexico

On the morning of February 19, Firefighter Hays and another firefighter were in the fire station conducting a work detail. They noticed a slight smell of propane in the fire station but could not find a source. Firefighter Hays reported the smell to two of the local propane suppliers and to the Rio Arriba County Fire Marshal's Office.

Based on the fire marshal's investigation of the explosion, Firefighter Hays arrived at the fire station at approximately noon. When he opened the door, he smelled gas. He opened the west apparatus bay door to attempt to clear the building of gas and began to shut off electrical service to the building. An explosion occurred and mortally injured Firefighter Hays. Other firefighters responding to the noise and smoke of the explosion found Firefighter Hays in the debris of the building and provided aid. Firefighter Hays was transported by ambulance to a meeting site where he was to be transported by medical helicopter. He died in the ambulance. The cause of death cited by the coroner's report was multiple system trauma.

The New Mexico State Fire Marshal conducted an independent investigation of this incident. The cause of the explosion was listed as accidental. Snow accumulation on the fire station from recent storms put pressure on a propane pipe leading into the fire station. A break occurred in the piping allowing propane to leak into the station. The propane was ignited accidentally, probably as Firefighter Hays cut electrical service to the building.

The Brazos Canyon Volunteer Fire Department (http://www.brazoscanyonfire.org/) has since finished building a new fire station. The new station was dedicated July 4, 2009, and was named the Michael J. Hayes Memorial Station in honor of Firefighter Hays (who had stepped down as Chief just a week before the incident).

February 21, 2008–1030 hrs
Joseph Eugene "Joey" Turner, Firefighter
Age 48, Volunteer

Homerville/Clinch County Volunteer Fire Department, Georgia

Firefighter Turner was participating in training at the Georgia Fire Academy. The training involved crawling through a maze wearing a SCBA, depleting the air supply in the SCBA, removing the facepiece, and exiting the maze.

When Firefighter Turner emerged from the maze, he suddenly collapsed. He was treated by other firefighters and an ambulance was dispatched to the scene. Paramedic-level care was provided and he was transported to the hospital.

Firefighter Turner was transferred to a regional hospital where he died the next day. The cause of death was a heart attack.

For additional information regarding this incident, please refer to NIOSH Fire Fighter Fatality Investigation and Prevention Program Report F2008-04 (http://www.cdc.gov/niosh/fire/reports/face200804.html).

February 23, 2008–0617 hrs
Shane Michael Stewart, Captain
Age 33, Volunteer

Ault-Pierce Fire Department, Colorado

Captain Stewart was the driver and only occupant of a 2,000-gallon engine-tanker responding to an emergency medical incident. During the response, he lost control of the vehicle, went off of the left side of the road, overcorrected, and went off the right side of the road. The apparatus rolled and ended up on its roof. Captain Stewart was ejected and died of traumatic injuries.

A report from the Ault-Pierce Fire Department noted that, since he was alone in the apparatus, Captain Stewart would have had to remove his seatbelt in order to reach the radio controls in the cab of the apparatus.

February 29, 2008–1604 hrs
Roger Dennis, Jr., Fire Apparatus Operator
Age 55, Career

San Antonio Fire Department, Texas

Fire Apparatus Operator Dennis was called in to work an overtime shift on February 29, 2008. When he arrived at the fire station, he and another firefighter began to ready their ambulance for service. The other firefighter noticed that Fire Apparatus Operator Dennis was taking his own blood pressure and asked if he was feeling well. Fire Apparatus Operator Dennis said that he was fine and they continued to prepare the ambulance.

Just over an hour later, at 1455 hours, Fire Apparatus Operator Dennis called his shift commander and told him that he was too sick to complete the shift. Another firefighter was called to replace him.

When the relief firefighter arrived at the station, he found that Fire Apparatus Operator Dennis was exhibiting signs of a cardiac event. He was transported to the hospital but was pronounced dead at 1815 hours. The cause of death was listed as blood clots in the lungs.

February 29, 2008–0600 hrs
Bradley Paul Holmes, Firefighter
Age 21, Volunteer

Pine Township Engine Company, Pennsylvania

Firefighter Holmes and the members of his fire department were dispatched to a fire in a duplex. Information provided to the firefighters on the scene confirmed that a victim was trapped in the building. Firefighter Holmes and another firefighter were ordered to conduct a search of an uninvolved area of the building.

Continued on next page

Water supply to the fire scene was interrupted due to frozen hydrants and the IC ordered an evacuation of the building. The fire intensified and trapped Firefighter Holmes and the other firefighter on the second floor of the building. A distress call was received from Firefighter Holmes and a Rapid Intervention Crew (RIC) was sent in to get the trapped firefighters.

One firefighter removed himself from the structure and Firefighter Holmes was found and removed by the RIC. Firefighter Holmes was transported to a local hospital and then transferred to a regional burn unit.

Firefighter Holmes died of complications of his injuries on March 5, 2008. He suffered burns on over 70 percent of his body. A resident of the home also died in the incident.

For additional information regarding this incident, please refer to NIOSH Fire Fighter Fatality Investigation and Prevention Program Report F2008-06 (http://www.cdc.gov/niosh/fire/reports/face200806.html).

March 3, 2008–1220 hrs
Rafael Vazquez, Fire-Rescue Lieutenant
Age 42, Career
Palm Beach County Fire-Rescue Department, Florida

Lieutenant Vasquez was attending a 40-hour supervisory class hosted by Palm Beach County Fire-Rescue. He was out of service for lunch at a fast food restaurant.

A gunman came out of the restroom area of the restaurant and fired randomly at patrons and workers in the business. Lieutenant Vasquez was struck by gunfire and died at the scene. Four other customers were injured and the gunman took his own life.

March 5, 2008–0830 hrs
Nicholas V. Picozzi, II, Lieutenant
Age 35, Volunteer
Lower Chichester Volunteer Fire Company, Pennsylvania

Lieutenant Picozzi and the members of his fire department responded to a structure fire that may have been the result of a tree falling on powerlines.

Firefighters advanced hoselines into the first story and basement of the residence. Lieutenant Picozzi was on the nozzle of the line that was advanced into the basement. He and another firefighter advanced to the bottom of the stairs and controlled all visible fire. Despite their efforts, heat conditions in the basement were severe and the firefighters decided to go back up the stairs. The firefighter backing up Lieutenant Picozzi was able to escape the basement via another route after encountering a blocked door, but Lieutenant Picozzi was unable to escape.

Lieutenant Picozzi was removed from the basement and transported to the hospital. He was pronounced dead at 1035 hours. He suffered third degree burns over 40 percent of his body. The carboxyhemoglobin level in his blood at autopsy was 50 percent.

March 7, 2008–0700 hrs
Victor Anthony Isler, Fire Control Specialist I
Age 40, Career
Salisbury Fire Department #1, North Carolina

Justin Edward Monroe, Fire Control Specialist I
Age 19, Part-Time (Paid)
Salisbury Fire Department #1, North Carolina

Salisbury Fire Department units were dispatched to a report of a fire in a large manufacturing occupancy. Business had been conducted on the site since 1939 and several major fires had occurred previously, the most recent in 1959. The building had been modified over time and totaled approximately 79,000 square feet of space. The building was one story with a partial basement. The business manufactured cabinets and architectural millwork on the first floor and in the basement.
Firefighters discovered a working fire in the office area in the basement of the building upon their arrival. Soon after the arrival of the onduty battalion chief, a third alarm was ordered.

Over the next hour, firefighters attempted to control the fire in the office areas in the basement and the first floor of the building and block extension of the fire. When these efforts proved unsuccessful, the fire strategy was changed to defensive.

Firefighters were allowed to enter the structure to attempt to prevent extension from the office areas and knock down any visible fire. Conditions again deteriorated and crews were pulled from the building. Once a new action plan was developed, crews were allowed back into the building. Fire conditions worsened again and firefighters were pulled from the building.

Firefighters, including Fire Control Specialists Isler and Monroe, were allowed to return to the structure as a part of a four-person crew to control extension from the basement office area and to attempt fire control. Fire conditions changed rapidly due to the collapse of an interior wall and the crew had to retreat, however, Isler and Monroe became trapped. Fire Control Specialist Isler was removed from the structure by RIC firefighters a short time later. The recovery of Fire Control Specialist Monroe was delayed by fire conditions.

Fire Control Specialist Isler was treated by firefighters and transported to the hospital. When he was recovered, Fire Control Specialist Monroe was obviously deceased.

The cause of death for Fire Control Specialist Isler was listed as burns. Upon his arrival at the hospital, his carboxyhemoglobin level was 21.6 percent. The cause of death for Fire Control Specialist Monroe was burns.

The Salisbury Fire Department has prepared an exhaustive analysis of this incident. The report is available at https://www.salisburync.gov/IncidentReport/index.html

For additional information regarding this incident, please refer to NIOSH Fire Fighter Fatality Investigation and Prevention Program report F2008-07 (http://www.cdc.gov/niosh/fire/reports/face200807.html).

March 9, 2008–1505 hrs
Raymond Barrett, Sr., Firefighter
Age 62, Volunteer

West Milford Township Fire Department, Apshawa Volunteer Fire Company #1, New Jersey

Firefighter Barrett was a member of the first engine company crew to arrive at the scene of a working fire in a residence. Firefighter Barrett and his crew advanced a 200-foot preconnected 1-3/4-inch handline to the front door of the structure. He used a flat headed axe to force open the door of the building.

Once the door was opened, Firefighter Barrett and two other firefighters advanced the hoseline into the structure and extinguished a kitchen fire. When the fire was knocked down, Firefighter Barrett and other firefighters began to open windows to ventilate the building.

Firefighter Barrett suddenly collapsed. Other firefighters in the immediate area called a mayday and removed Firefighter Barrett from the structure. Emergency medical personnel on the scene began CPR immediately and he was transported to the hospital.

Despite all efforts, Firefighter Barrett did not recover and was pronounced dead at the hospital. The cause of death was a heart attack.

March 16, 2008–0925 hrs
Walter William Michl, Firefighter
Age 76, Volunteer

Roanoke-Wildwood Volunteer Fire Department, North Carolina

At 0925 hours on March 16, 2008, Roanoke-Wildwood firefighters were dispatched to a reported furnace fire in a manufactured home. As firefighters responded, the home was reported as fully involved.

Firefighter Michl rose from bed and responded to the incident. When his wife went to the kitchen in their home approximately 90 minutes later, she found Firefighter Michl in a chair, unresponsive and not breathing. EMS personnel responded to the home and pronounced Firefighter Michl dead. The cause of death was a heart attack.

March 17, 2008–1230 hrs
Terrance Dale Crockett, Firefighter
Age 48, Career

Kansas City Fire Department, Missouri

Kansas City firefighters were dispatched to a report of a fire in a vacant residence. Arriving firefighters found a working fire. The fire was controlled and firefighters began to overhaul the interior of the structure.

Continued on next page

Firefighter Crockett was working on the interior of the building, wetting down hot spots. He was wearing his full structural protective clothing equipment as well as a SCBA. He was breathing air from the SCBA.

Firefighter Crockett suddenly collapsed and was immediately removed from the building by other firefighters. The crew of a paramedic ambulance on the scene began treatment immediately and he was transported to the hospital. The total time elapsed from his collapse to his arrival at the emergency room was 8 minutes. Despite this rapid delivery to a definitive care facility, Firefighter Crockett did not survive. The cause of death was listed as a heart attack.

The fire was intentionally set.

March 17, 2008–1420 hrs
John Patrick Delaney, Captain
Age 44, Career
Mesa Fire Department, Arizona

Captain Delaney was on duty in his position in the administrative offices of the Mesa Fire Department's EMS office. He had been on light duty receiving kidney dialysis and was on a waiting list for a donor kidney.

Captain Delaney returned from lunch and was found unconscious less than 5 minutes later by a coworker. CPR was begun immediately and he was transported to the hospital. He was not revived. The cause of death was listed as a heart attack.

March 21, 2008–2116 hrs
Donald Lee Grubor, Deputy Fire Chief
Age 42, Volunteer
Lewistown Fire Protection District, Illinois

Chief Grubor responded in a fire department vehicle to a mutual-aid structure fire in a residence. Chief Grubor, wearing full structural protective clothing, directed the work of his department's tanker (tender) in a water supply operation.

Once the fire was extinguished, Chief Grubor assisted the fire investigator by moving and shoveling debris so that the cause and origin of the fire could be determined. Chief Grubor was on the scene for over an hour and then returned to the fire station to return equipment to service.

Just after midnight on March 22, 2008, Chief Grubor headed home. It is estimated that he arrived home at 0030 hours. He showered and went to bed after complaining to his wife about pain in the right side of his head.

At 0139 hours, the Lewistown Fire Protection District was dispatched to an emergency medical incident. Chief Grubor did not respond.

Continued on next page

At 0150 hours, fire department and emergency units were dispatched to Chief Grubor's home. Responders found him in distress and transported him to a local hospital. He was pronounced dead at 0253 hours. His death was caused by a heart attack.

March 26, 2008–1357 hrs
Brent Allen Lovrien, Firefighter
Age 35, Career

Los Angeles City Fire Department, California

Firefighter Lovrien and the members of his engine company were dispatched along with other fire department units to investigate smoke coming from a structure in the area of 8800 South Sepulveda Boulevard. The incident occurred in the same area where another engine was assigned to an electrical vault fire in the street.

Upon their arrival on the scene, Firefighter Lovrien and another firefighter were assigned to force entry into a room that housed electrical equipment and had been emitting smoke. Firefighter Lovrien used a rotary saw to attempt to gain access. Sparks from the saw ignited combustible gases that had accumulated in the locked room and a large explosion occurred. He was mortally injured in the explosion when he was struck both by the rotary saw and the door.

Firefighter Lovrien was transported to the hospital by fire department ambulance. Despite rapid treatment and transportation, he did not survive. The cause of death was listed as blunt force injuries to the head.

March 28, 2008–0930 hrs
Eric DeWayne Speed, Firefighter
Age 33, Volunteer

Caddo Parish Fire District Two, Louisiana

Firefighter Speed was responding as the driver of a 3,000 gallon water tanker (tender) to a structure fire in a neighboring fire district. As it responded, the vehicle entered a 90-degree curve with a posted cautionary speed of 10 miles per hour. The apparatus failed to negotiate the turn and left the roadway. The apparatus rolled onto the driver's side and struck a large pine tree.

Firefighter Speed was trapped in the apparatus. Arriving firefighters were unable to extricate him until a tow truck was used to pull the truck off of the tree. Firefighter Speed was extricated approximately 45 minutes after the crash and flown by medical helicopter to the hospital. He was pronounced dead at the hospital.

Firefighter Speed was wearing his seatbelt at the time of the crash. Excessive speed was cited in the law enforcement report as a factor in the crash.

For additional information regarding this incident, please refer to NIOSH Fire Fighter Fatality Investigation and Prevention Program Report F2008-10 (http://www.cdc.gov/niosh/fire/reports/face200810.html).

April 4, 2008–0610 hrs
Robin Marie Zang-Broxterman, Captain
Age 37, Career
Colerain Township Department of Fire & EMS, Ohio

Brian William Schira, Firefighter-EMT
Age 29, Part-Time (Paid)
Colerain Township Department of Fire & EMS, Ohio

Captain Broxterman and Firefighter Schira were assigned to Engine 102 along with a fire apparatus operator and another firefighter. Their unit was dispatched along with other firefighters to the report of a fire in a residence. Engine 102 was the first fire department unit on the scene and laid a supply line up the extended driveway to the residence. Captain Broxterman reported moderate smoke showing and established Command at 0623 hours. While donning her protective clothing, Captain Broxterman was advised by her driver that the home's resident said that the fire was in the basement.

Firefighter Schira advanced a 150-foot, 1-3/4-inch handline to the front door of the structure. Captain Broxterman and Firefighter Schira entered the structure with a dry handline and called for the line to be charged. Engine 102's other firefighter entered the interior after checking the deployment of the supply line.

At 0627 hours, Captain Broxterman radioed that Engine 102 was making entry into the basement and reported heavy smoke. After a request for water, the handline was charged at approximately 0629 hours.

At 0634 hours, the second firefighter from Engine 102 told another officer that he could not find his crew. The officer reported this fact to Command and Mayday operations were initiated. A second alarm was requested and a RIC was deployed.

Captain Broxterman was located at 0708 hours and Firefighter Schira was located at 0729 hours. Both firefighters were buried under collapsed structural components and were declared dead at the scene. The cause of death for both firefighters was smoke inhalation.

The committee that investigated the fire believes that one or more catastrophic events occurred inside of the structure between 0630 and 0634 hours. A large area of the first floor collapsed into the basement.

A preliminary report on this incident has been prepared by the Colerain Township Department of Fire & EMS. The report is available at the fire department Web site (http://www.coleraintwp.org/fire.cfm).

For additional information regarding this incident, please refer to NIOSH Fire Fighter Fatality Investigation and Prevention Program report F2008-09 (http://www.cdc.gov/niosh/fire/reports/face200809.html).

53

April 8, 2008–1545 hrs
Michael David Crotty, Deputy Chief
Age 24, Volunteer
Lawrence Park Township Volunteer Fire Department, Pennsylvania

Chief Crotty was in Command of an exterior fire at a plastics manufacturing facility. The aerial ladder on a quint apparatus was set up to help with control of the fire. When the extended aerial ladder pipe was pressurized, the motorized water monitor and 30 feet of aluminum pipe were projected out the end of the ladder. The projected assembly struck Chief Crotty and he sustained fatal injuries.

The type of aerial ladder installed on this quint apparatus was capable of operating in a water tower or rescue mode. In the water tower mode, the shuttle that carried the water monitor and pipe was pinned to the fly section of the ladder to maximize the operating height of the water monitor. In the rescue mode, the shuttle was pinned to a lower ladder section so that it would not interfere with the placement of the ladder for rescue purposes. At this incident, the shuttle was not properly secured and was projected off of the ladder when the ladder's water pipe was pressurized.

For additional information regarding this incident, please refer to NIOSH Fire Fighter Fatality Investigation and Prevention Program Report F2008-12 (http://www.cdc.gov/niosh/fire/reports/face200812.html).

April 8, 2008–2034hrs
Rickey Stephens Morris, Firefighter
Age 54, Career
Sedalia Fire Department, Missouri

Firefighter Morris was assigned to the crew of Truck 1 for the shift that began at 0800 hours on April 8. At 2031 hours, Truck 1 and two engine companies were dispatched to a report of a fire in a residence. Truck 1 was the first fire department unit on the scene. The CO reported smoke visible and Command was established.

Firefighter Morris and another firefighter advanced a charged 1-3/4-inch preconnected handline into the structure. The firefighters had difficulty locating the fire and smoke and heat conditions intensified. Firefighter Morris and the other firefighter became separated.

Rapid fire progress occurred and firefighters heard the sounding of a Personal Alert Safety System (PASS) device inside of the structure. Firefighter Morris was discovered and removed from the structure.

Firefighter Morris was treated on the scene and flown by medical helicopter to a regional trauma facility. He suffered second and third degree burns over 43 percent of his body and respiratory burns.

Firefighter Morris died as the result of his injuries on April 17, 2008.

April 12, 2008–0601 hrs
Charles Carter Fraley, Jr., Fire Chief
Age 65, Volunteer

Macon Fire Department, Mississippi

Chief Fraley and the members of his fire department responded to a report of a structure fire at a local vocational school. Chief Fraley responded directly to the scene and found a working fire.

Chief Fraley directed the deployment of firefighters and hoselines at the scene and became ill. He asked his assistant chief to take Command of the firefight. Chief Fraley asked a local police officer to drive him to the hospital. During the drive, he complained of shortness of breath. Upon his arrival at the hospital, he walked into the emergency room and was treated by medical personnel.

Chief Fraley's condition deteriorated and he was pronounced dead at 0730 hours. The cause of death was listed as a heart attack.

April 15, 2008–1545 hrs
Terry W. DeVore, Fire Chief
Age 30, Volunteer

Olney Springs Volunteer Fire Department, Colorado

John Wesley Schwartz, Jr., Firefighter
Age 38, Volunteer

Olney Springs Volunteer Fire Department, Colorado

Chief Devore was the driver and Firefighter Schwartz was the front seat passenger in a 1988 GMC brush truck. The unit was responding to the Ordway wildland fire. Conditions in the area included blowing dust, smoke, and wind.

During the response, the crew attempted to drive over a wooden bridge over a ravine. Unbeknownst to the firefighters, the bridge had been damaged by fire and failed as the brush truck drove over it.

The truck came to rest against the embankment at the opposite side of the ravine and caught fire. According to the county coroner, both firefighters died of traumatic injuries in the crash and subsequently burned. The firefighter's remains were recovered the next day when fire conditions permitted.

April 15, 2008–1820 hrs
Gert H. "Jerry" Marais, Pilot
Age 42, Wildland Contract

Aero Applicators under contract with the Colorado State Forest Service, Colorado

Pilot Morris was operating an AirTractor 602 single engine air tanker. He had initially declined the mission due to weather and terrain conditions, but then agreed to proceed to the scene to evaluate conditions.

Upon arriving at the fire, the pilot tried a dry run in the area where a drop had been requested. The winds and turbulence were too strong so another drop location was designated. Pilot Morris made the drop as requested, but the aircraft stalled and crashed.

The National Transportation Safety Board (NTSB) determined that the probable cause(s) of this accident were as follows: The pilot's failure to maintain aircraft control following the jettison of the load during an aerial firefighting mission, which resulted in an inadvertent stall and impact with terrain. Contributing to the accident were the improperly configured aircraft for the flight, the gusty wind conditions, and the pressure to complete the mission.

For additional information about this crash, consult the NTSB Web site at (http://www.ntsb.gov/ntsb/query.asp - NTSB identification DEN08GA076).

April 20, 2008–2100 hrs
Riley Joseph Terrebonne, Jr., Firefighter
Age 29, Volunteer

Springfield Volunteer Fire and Rescue Department–Livingston Parish Fire Protection District 2, Louisiana

Firefighter Terrebonne stopped at the roadside to provide assistance at the scene of a vehicle crash. As he worked on the scene, he was struck by another vehicle that entered the crash scene.

Firefighter Terrebonne was flown to a local hospital by air ambulance, stabilized, and then taken to a regional trauma facility. He did not recover and was pronounced dead on April 21 at 1230 hours.

April 24, 2008–1937 hrs
Gary D. Remling, Firefighter
Age 56, Volunteer

Belltown Fire Department, Connecticut

Firefighter Remling and the members of his fire department responded to a mutual-aid structure fire in a neighboring fire district. He was assigned to a standby crew that was providing coverage for several fire districts that were involved in the structure fire incident. When the Belltown engine returned to the fire station, Firefighter Remling assisted with returning the unit to service.

Continued on next page

The next day, at 1204 hours, Stamford firefighters responded to a report of an unconscious person in a car in a residential driveway. Crews discovered Firefighter Remling in his personal vehicle in cardiac arrest. He was transported to the hospital but did not survive.

April 29, 2008–1126 hrs
Jeremy L. Jylka, Firefighter
Age 34, Paid-on-Call
Pine City Fire Department, Minnesota

Firefighter Jylka had just completed his first year of service with the Pine City Fire Department. The department was paged to respond to a grass fire. Firefighter Jylka responded to the fire station, donned his personal protective clothing, and was directed by the Captain in charge to ride in the passenger seat of the department's grass truck.

As the apparatus proceeded to the scene, the driver of the truck noticed that Firefighter Jylka was gasping for air and was in obvious distress. The driver requested an ambulance and returned to the fire station.

Firefighter Jylka was treated and transported but did not survive. The cause of death was listed as a heart attack.

May 10, 2008–1750 hours
Tyler Heath Casey, Firefighter/First Responder
Age 21, Volunteer
Seneca Area Fire Protection District, Missouri

On May 10, 2008, at approximately 1730 hours, the Seneca Area Fire Protection District was dispatched by Newton County Central Dispatch for storm spotting duties. Firefighter Casey responded to the dispatch and was assigned by the IC to a post near the intersection of Highway 43 and Iris Road. Newton County Emergency Management had received information that a storm was active in that area.

The storm behaved unpredictably and began to move in Firefighter Casey's direction at approximately 70 miles per hour. Other units posted in the area began to report damage and a distress message was transmitted by radio from Firefighter Casey. The storm had overrun his position and was right on top of him when he made his last radio transmission at approximately 1750 hours. Firefighter Casey's report resulted in the sounding of tornado sirens in his county and the declaration of a tornado warning in his county and several surrounding counties.

Reports of storm damage in the area of Highway 43 and Iris Road began to stream into the Newton County Central Dispatch Center. First-arriving units found Firefighter Casey trapped in his vehicle suffering from severe trauma to his head and chest. Firefighter Casey was extricated from his vehicle by mutual-aid firefighters and then transported by ambulance to Freeman West Hospital in Joplin, Missouri. He died of his injuries on May 12, 2008, at 1352 hours.

May 12, 2008–1230 hrs
Joseph Randall "Randy" Mixon, Training Captain
Age 51, Career

Birmingham Fire and Rescue Service Department, Alabama

Captain Mixon was on duty the morning of May 12, when he took ill and left for home at approximately 1230 hours.

Captain Mixon went to the doctor with upper respiratory-type symptoms and returned to his residence after being examined. He suffered a cerebrobascular accident (CVA) at approximately 0030 hours on May 13 and was transported to the hospital. He passed away later in the day.

May 19, 2008–1555 hrs
Raymond F. "Rocky" Eusden, Fire Chief
Age 57, Volunteer

Aston-Beechwood Volunteer Fire Company, Pennsylvania

Chief Eusden was assisting with apparatus and equipment maintenance duties in his fire station. He attempted to open a compartment door on the department's heavy rescue truck but was unable. He complained of numbness in his arm and his speech became slurred.

Chief Eusden died on May 20, 2008, as a result of a CVA.

May 23, 2008–2300 hrs
Jay C. Maddy, Firefighter
Age 41, Volunteer

Eaton Volunteer Fire Department, Indiana

Firefighter Maddy and members of his fire department had responded to numerous arson fires since Wednesday evening, May 21, 2008. There had been 21 incidents in less than 24 hours. A curfew was imposed and a "fire watch" was established. Firefighter Maddy served on a fire watch detail until approximately 0200 hours on May 23, 2008. Several suspects were arrested.

After being at home for a short period of time, he assisted with a fire department fundraiser work detail that started at 0800 hours and was concluded after noon.
At approximately 2300 hours, Firefighter Maddy was taken to a hospital with chest pain. He died a few hours later in surgery.

May 24, 2008–Time Unknown
Richard A. "Rick" Burns, Firefighter
Age 43, Career

Pittsburgh Bureau of Fire, Pennsylvania

Firefighter Burns suffered a fatal heart attack while on duty in his fire station.

May 31, 2008–1845 hrs
Thomas Russell Topping, Firefighter
Age 28, Volunteer

Barnsdall Rural Fire Department, Oklahoma

Firefighter Topping was enrolled in a Firefighter I training program. The class met twice a week and on Saturdays for 8 weeks in Dewey, Oklahoma. At the completion of the course, firefighters must complete a 2-day practical and written certification test. Firefighter Topping and other firefighters travelled to the State fire training organization's facility near Stillwater, Oklahoma for their test.

The day began with a briefing about the events that would occur on the first day of testing. The outside temperature was 75 °F (241 °C) with 69-percent humidity. This was the hottest day to this point in the year.

Firefighters participated in a series of training activities that morning that included work in a burn building, fire extinguisher training, and car fire, pallet, and dumpster fire props. Most firefighters remained in their turnout gear throughout the exercises. Water in 5-gallon cooler cans was provided for firefighters as well as some areas of shade.

At approximately 1100 hours, a firefighter was transported by ambulance from the training site for a heat-related illness. The class broke for lunch at 1200 hours. The break was approximately 90 minutes.

After the break, students rejoined their instructors and continued with the training program. The outside temperature had reached 85 °F (291 °C) by this point with humidity of 65 percent and winds at 15 mph. By 1540 hours, the highest temperature of the day was reached at 91 °F (33 °C), humidity of 52 percent, and winds at 18 mph.

Firefighter Topping and his group participated in a rotation in the burn building that lasted approximately 2-1/2 hours. At the completion of the burn building rotation, Firefighter Topping and his group were rotated to another prop. Firefighters were exhibiting signs of fatigue. At the conclusion of the exercises for the day, Firefighter Topping asked for and received permission to go to the restroom. He seemed in good health and good spirits.

At approximately 1815 hours, Firefighter Topping was discovered unconscious in the restroom. He was removed from the restroom and CPR was initiated. An AED was applied and he was transported to the hospital by fire department ambulance. Firefighter Topping was pronounced dead at the hospital at 1911 hours.

The cause of death was listed as hyperthermia and dehydration. Firefighter Topping had an intestinal condition that may have contributed to his illness.

June 6, 2008–1610 hrs
Rufus Edwin Brinson, Jr., Firefighter
Age 50, Volunteer

Reelsboro Fire Department, North Carolina

Firefighter Brinson was instructing a Firefighter I class at Pamlico Community College. He provided instruction in the classroom for the morning session from 0800 hours to 1145 hours.

After lunch, Firefighter Brinson and other firefighters picked up a pumper at their fire station and made preparations for smokehouse exercises. Firefighter Brinson operated the pump panel and led students through a flammable liquids fire scenario. The weather was very hot.

He complained of not feeling well and another instructor led the last group of students through the flammable liquids prop. The balance of the afternoon's outside activities were cancelled due to the heat and the fact that Firefighter Brinson was not feeling well.

Firefighter Brinson assisted with cleaning the training facility and reloading hose on the pumper. As the apparatus was filled with water, he collapsed and was unresponsive. CPR was initiated and Firefighter Brinson was transported to the hospital. Firefighter Brinson did not recover and died due to a heart attack.

June 14, 2008–0530 hrs
Colin Gene Thomas, Second Assistant Chief
Age 51, Volunteer

Verona Volunteer Fire Department, North Carolina

A fire in an inaccessible area of a military base firing range had been burning for over 2 months. The fire was contained by means of a firebreak line that had been created using heavy equipment. On the morning of June 14, smoke from the fire mixed with fog and created zero visibility conditions on a highway near the military base.

A series of vehicle crashes occurred in the smoky area and the Verona Volunteer Fire Department was dispatched to assist. Chief Thomas arrived on the first engine company and reported heavy smoke and poor visibility conditions.

As law enforcement officials arrived on the scene, additional crashes occurred, at least one involving a law enforcement vehicle. The roadway was in the process of being shut down by law enforcement officers.

Chief Thomas was outside of the engine, wearing full turnout gear and a reflective vest, when he was struck by a tractor-trailer truck that entered the scene. A sheriff's deputy was struck and killed and another deputy was injured. The report on the incident estimated the speed of the tractor-trailer at 55 mph when it entered the scene.

Chief Thomas died of multiple blunt trauma.

For additional information regarding this incident, please refer to NIOSH Fire Fighter Fatality Investigation and Prevention Program Report F2008-17 (http://www.cdc.gov/niosh/fire/reports/face200817.html).

June 16, 2008–1215 hrs
Kevin Patrick Pryor, Firefighter
Age 31, Career

Newport Beach Fire Department, California

Firefighter Pryor was discovered by a friend at his residence having fallen ill just several hours after he returned from deployment on the department Strike Team working the Humboldt Fire in Northern California. Pryor was transported to the hospital but passed away the next day from a nontraumatic brain hemorrhage.

June 20, 2008–0130 hrs
Gary Lawrence Studer, Captain
Age 61, Career

Whitehouse Fire Department, Ohio

Captain Studer was assigned for the shift to Life Squad 9. He and his crew performed station and apparatus maintenance duties in the morning. The unit responded to a motorcycle crash at 0933 hours. On the scene, they found a motorcycle driver sitting alongside the road with his motorcycle. The driver was bleeding and confused. Captain Studer and other firefighters provided treatment and the driver was flown by medical helicopter to the hospital.

At 0130 hours, Captain Studer's unit was dispatched to an emergency medical incident. When his partner shook him to respond on the call, he found that Captain Studer was unresponsive. The firefighter called for assistance and Captain Studer was treated and transported to the hospital.

Captain Studer died as the result of a CVA on June 28, 2008.

June 21, 2008–0800 hrs
George E. Crocker, Fire Chief
Age 32, Volunteer

Pine Level Volunteer Fire Department, North Carolina

Chief Crocker responded to a fire and three EMS incidents in his role as the fire chief for the Pine Level Volunteer Fire Department. The next morning, he was found dead in his home by another firefighter. His death was caused by a heart attack.

Chief Crocker was also a career firefighter with the Raleigh Fire Department.

June 26, 2008–1220 hrs
Jeff Powers, Deputy Chief
Age 44, Career
Southern Marin Fire Protection District, California

Chief Powers was talking with firefighters in a fire station when he suddenly collapsed. Firefighters treated Chief Powers and transported him to the hospital but he did not survive the heart attack.

June 29, 2008–1547 hrs
Michael James MacDonald, Firefighter
Age 26, Wildland Part-Time
Chief Mountain Interagency Hotshot Crew–Blackfeet Tribe, Montana

On June 27, 2008, Firefighter MacDonald was battling a fire on the north rim of the Grand Canyon National Park when he was bitten by an insect and taken to a nearby hospital for treatment. On Sunday, June 29, 2008, he suffered anaphylactic shock from the antibiotics being used to treat the insect bite and was being flown by medical helicopter to the Flagstaff Medical Center.

Approximately 1/4-mile from the hospital, the helicopter collided with another medical-transport helicopter that was also on approach to the hospital. The crash killed Firefighter MacDonald and six other people in the two aircrafts.

For additional information about this crash, consult the NTSB Web site at (http://www.ntsb.gov/ntsb/query.asp - NTSB identification DEN08MA116).

July 2, 2008–1200 hrs
Robert Roland, Firefighter
Age 63, Volunteer
Anderson Valley Volunteer Fire Department, California

Firefighter Roland was assigned the position of lookout on the Oso Fire on July 2, 2008. He saw other firefighters carrying hose to progress a hose-lay and started to help them carry hose up a hill when he experienced extreme fatigue and respiratory distress.

Initially it was thought to be heat-related distress. He was moved from the fireground in an air-conditioned personal vehicle to meet the ambulance which transported him to the hospital. He was initially treated in the emergency room and then transferred to the intensive care unit (ICU) where he died the next morning, July 3, 2008, at approximately 0400 hours from a massive heart attack.

July 2, 2008–1241 hrs
Joe Pat Jordan, Firefighter
Age 71, Volunteer

Pickton-Pine Forest Volunteer Fire Department, Texas

Firefighter Jordan was responding as a passenger in a fire apparatus to a reported motor vehicle crash with entrapment when he went into cardiac arrest. The apparatus was pulled to the side of the road and CPR was initiated on Firefighter Jordan and an AED was applied.

He was revived and transported to the local hospital. He was transferred by air ambulance to a regional hospital. He never regained consciousness. He was removed from life support and succumbed to his injuries at 2112 hours on July 4, 2008.

The Texas State Fire Marshal's Office prepared a detailed report on this incident. The report is available at (http://www.tdi.state.tx.us/fire/fmloddinvesti.html).

July 5, 2008–1747 hrs
Robert Leland Knight, Fire Chief
Age 42, Volunteer

Teague Volunteer Fire Department, Texas

Chief Knight and the members of his fire department were dispatched to a report of a structure fire in a commercial occupancy. Chief Knight responded to the scene in a tanker (tender) apparatus from the scene of a wildland fire where he had been working. He arrived approximately 10 minutes after dispatch and found a working fire in an automotive repair and upholstery shop.

The involved structure was a wood-frame building with metal siding and roof. The building measured approximately 40 feet by 140 feet. The south, front end of the building had a brick façade that rose to a peak at the roof to a height of 20 feet.

Chief Knight was operating a nozzle at a doorway of the south end of the building when the two-story brick façade collapsed outward, pinning him as he was running away. Chief Knight was immediately extricated by fellow firefighters and civilian witnesses and treated by the onscene medic unit. He was transported by ground ambulance to the local helipad and flown by air ambulance to the East Texas Medical Center in Tyler, Texas. Chief Knight succumbed to extensive traumatic injuries and died at 2138 hours on July 5, 2008.

The Texas State Fire Marshal's Office prepared a detailed report on this incident. The report is available at (http://www.tdi.state.tx.us/fire/fmloddinvesti.html).

July 7, 2008–2008 hrs
Richard L. Kear, Firefighter
Age 58, Paid-on-Call
Pitt Township Volunteer Fire Department, Ohio

Pitt Township firefighters were initially dispatched to a vehicle fire that was later updated to a garage fire. Firefighter Kear responded to the fire station and got into the driver's seat of Engine 10, a 1991 commercial chassis pumper with a 1,000-gallon water tank.

During the response, Firefighter Kear came upon a farm tractor with implements approaching from the opposite direction. The tractor pulled as far to the side of the road as possible and stopped. Firefighter Kear pulled the pumper to the right and passenger side wheels of the apparatus left the roadway. He steered left to bring the truck back on the road and then steered to the right when the apparatus jumped toward the opposing lanes. The rear of the apparatus came around and the apparatus rolled several times.

Firefighter Kear was not wearing his seatbelt and was ejected during the crash. A front seat passenger in the apparatus was also not wearing a seatbelt but remained inside of the vehicle.

Firefighter Kear suffered fatal traumatic injuries. The passenger firefighter received nonlife threatening injuries. The speed of the apparatus prior to the crash was estimated by a witness to be 55 to 60 mph.

July 8, 2008–1800 hrs
Ryan Thomas Barker, Firefighter
Age 25, Volunteer
West Hill Fire Department, New York

Firefighter Barker was the driver and sole occupant of a 1978 commercial chassis engine apparatus equipped with a 750-gallon tank. The apparatus was returning to the fire station from a fire response.

As the apparatus traveled down a winding, steep road, Firefighter Barker lost control of the vehicle. The apparatus went off the left side of the roadway, returned to the roadway, then went off the right side of the roadway, and overturned.

Firefighter Barker was not wearing a seatbelt and he was ejected from the apparatus during the crash. This response was Firefighter Barker's first emergency call driving this apparatus. The cause of death was listed as multiple trauma and crush injuries to the chest and abdomen.

For additional information regarding this incident, please refer to NIOSH Fire Fighter Fatality Investigation and Prevention Program Report F2008-25 (http://www.cdc.gov/niosh/fire/reports/face200825.html).

July 20, 2008–2314 hrs
David P. Meron, Sr., Fire Police Officer
Age 58, Volunteer
Hoosick Falls Fire Department, New York

Firefighter Meron responded to an alarm for a possible transformer malfunction. Once released from that alarm, he returned to the fire station. When firefighters returned from another alarm, they found Firefighter Meron unconscious and unresponsive in his vehicle at the entrance to the parking lot of the fire station. CPR was initiated by fire department personnel and Firefighter Meron was transported to a local medical center. He was pronounced dead at the hospital after all resuscitative efforts failed. The cause of death was reported to be a heart attack.

July 21, 2008–0545 hrs
Ryan Andrew Hummert, Firefighter
Age 22, Career
Maplewood Fire Department, Missouri

At 0540 hours, Maplewood firefighters were dispatched to a report of a vehicle fire. The units arrived at 0542 hours and reported a vehicle fire and said that they were using a booster line. At 0546 hours, firefighters reported that they were taking gunfire. Moments later, firefighters advised that a firefighter and a police officer were down and were still taking gunfire.

Firefighter Hummert suffered a gunshot wound to the head and was out of reach of the other firefighters on the scene. From a distance, he appeared to have died soon after being shot.

A gunman had apparently set the vehicle fire to draw responders into the scene. The gunman was barricaded in a single-family residence.

An armored law enforcement vehicle was brought to the scene and extricated the trapped police officers and firefighters from the scene. The armored vehicle was also used to recover Firefighter Hummert and bring him to an EMS unit at 0737 hours. Firefighter Hummert arrived at the hospital at 0800 hours and was pronounced dead.

In addition to Firefighter Hummert's death, two police officers were injured. The gunman also died in the incident.

July 22, 2008–1945 hrs
Brian James Munz, Firefighter
Age 24, Volunteer
Fairbury Fire Department, Illinois

A fire was reported in the basement of a residential structure. The first fire department unit on the scene made contact with the homeowner and confirmed that everyone was outside of the home. Initial arriving fire companies established a water supply and advanced a handline into the basement of the structure. The crew could not make access to the basement due to debris in the hallway and falling debris. Smoke from the structure was becoming darker and crews had not located the fire. A ventilation fan was placed to draw smoke from the basement. The smoke conditions continued to worsen.

Firefighter Munz responded as a part of an engine company crew as mutual aid. When his unit arrived on the scene, the IC directed them to enter the first floor of the structure to open windows and look for fire extension. At some point, firefighters determined that the floor was softening on the first floor and decided to evacuate the structure.

The first and second members of Firefighter Munz' crew made it out the front door and a collapse occurred. Firefighter Munz fell into the fire-involved basement of the structure. Evacuation horns were sounded and firefighters used multiple handlines to attempt to control the fire in Firefighter Munz' location.

Once the fire was knocked down sufficiently to allow entry, a ground ladder was lowered into the hole and firefighters retrieved Firefighter Munz.

Firefighter Munz had expired prior to being removed from the structure. During the autopsy, it was found that Firefighter Munz had first and second degree burns over 50 percent of his body. His carboxyhemoglobin level was normal. The cause of death was listed as positional asphyxiation; Firefighter Munz was crushed by debris, principally a couch, and was unable to breathe. His facepiece stayed in place.

July 23, 2008–0648 hrs
Frank L. Wichlacz, Fire Chief
Age 76, Volunteer
Pulaski Tri-County Fire Department, Inc., Wisconsin

Chief Wichlacz and other members of his fire department had just returned to their station after fighting a structure fire. Firefighters were in the process of returning equipment to service and refilling apparatus water tanks.

A firefighter's pickup truck was blocking one of the rear bay doors at the fire station. Another firefighter received permission from the owner to move the pickup. The firefighter intended to back the pickup away from the bay and relocate the pickup to another bay just to the right of the original bay to allow access. The firefighter got into the pickup wearing bunker pants and wet rubber boots.

Continued on next page

66

As the pickup was pulled forward, it unexpectedly accelerated and struck the bay door. Chief Wichlacz happened to be walking through the area at that moment and was crushed between the front of the pickup and the rear of a tanker (tender) parked in the bay.

Chief Wichlacz was quickly removed and transported by ambulance to the hospital. He was pronounced dead at the hospital. The cause of death was listed as blunt trauma to the chest.

For additional information regarding this incident, please refer to NIOSH Fire Fighter Fatality Investigation and Prevention Program Report F2008-27 (http://www.cdc.gov/niosh/fire/reports/face200827.html).

July 25, 2008–1350 hrs
Andrew Jackson Palmer, Firefighter
Age 18, Wildland Part-Time
Olympic National Park, Washington

Firefighter Palmer was assigned as a part of an engine crew to the Iron Complex fire on the Shasta-Trinity National Forest near Weaverville, California. When Firefighter Palmer and his crew arrived at the Incident Command Post (ICP), they could not go to work immediately due to a mechanical problem with their apparatus.

After working logistical assignments at the camp, Firefighter Palmer and his crew were assigned to a tree felling team. The team worked ahead of firefighters mopping up a fire that was under control. The felling team removed or dropped hazardous trees to allow other firefighters to work more safely.

At approximately 1350 hours, Firefighter Palmer was injured by a falling tree. Helicopter access to the site was prevented by smoke conditions so other firefighters carried Firefighter Palmer to a location where he was picked up by a USCG helicopter.

Firefighter Palmer went into cardiac arrest in the helicopter and was pronounced dead.

July 26, 2008–1530 hrs
Daniel Bruce Packer, Fire Chief
Age 49, Career
East Pierce Fire & Rescue, Washington

Chief Packer was dispatched as a Division Supervisor to the Siskiyou Complex of fires on the Klamath National Forest in Northern California. Chief Packer and other firefighters were assigned responsibility for the Panther Fire.

In order to prepare to assume management responsibility for the fire, Chief Packer and another division supervisor were assigned to scout the area. At approximately 1530 hours, the fire spread quickly and burned over Chief Packer's position.

Chief Packer deployed his fire shelter in an unsuccessful attempt to protect himself from the rapidly advancing flame front. A subsequent autopsy revealed that Chief Packer died of smoke inhalation and thermal burns.

August 3, 2008–1900 hrs
Gerald R. "Gerry" Leduc, Lieutenant
Age 50, Career

Tiverton Fire Department, Rhode Island

Firefighter Leduc responded to a report of a man who went overboard from a boat into Stafford pond. Leduc was part of a dive team who was searching for the lost subject when he suddenly had difficulty and went into cardiac arrest. He was rushed back to shore where despite all resuscitative efforts, he succumbed to his injury.

August 5, 2008–1930 hrs
Shawn Patrick Blazer, Firefighter
Age 30, Wildland Contract

Grayback Forestry, Inc., Oregon

Scott Albert Charlson, Firefighter
Age 25, Wildland Contract

Grayback Forestry, Inc., Oregon

Edrik Juan Gomez, Firefighter
Age 19, Wildland Contract

Grayback Forestry, Inc., Oregon

Matthew Aaron Hammer, Firefighter
Age 23, Wildland Contract

Grayback Forestry, Inc., Oregon

James N. Ramage, Pilot
Age 63, Wildland Full-Time

United States Forest Service, California

Steven Caleb Renno, Firefighter
Age 21, Wildland Contract

Grayback Forestry, Inc., Oregon

Bryan James Rich, Firefighter
Age 29, Wildland Contract

Grayback Forestry, Inc., Oregon

Continued on next page

Roark David Schwanenberg, Pilot
Age 54, Wildland Contract
Carson Helicopters, Inc., Oregon

David Elijah Steele, Firefighter
Age 19, Wildland Contract
Grayback Forestry, Inc., Oregon

A large number of firefighters were assigned to the Iron Complex fire in California's Trinity Alps Wilderness. A forecast for worsening weather necessitated the removal of approximately 50 firefighters from the area. A Sikorsky S61-N helicopter was assigned to ferry the firefighters to a safer location.

The helicopter completed two trips and had made a refueling stop. After it refueled, it returned for its third load of passengers. During departure, according to a NTSB preliminary report, the helicopter experienced a loss of power to the main rotor during takeoff, and subsequently impacted trees and terrain. The aircraft came to rest on its left side. A postcrash fire consumed the aircraft.

Nine firefighters, including two pilots died in the crash and a pilot and three firefighters were severely injured.

For additional information about this crash, consult the National Transportation Safety Board Web site at http://www.ntsb.gov

....

August 9, 2008–1030 hrs
Sean T. Whiten, Captain
Age 47, Volunteer
Roscoe Volunteer Fire Company, Pennsylvania

Captain Whiten had been assisting other instructors at a structural burn session at the Smithton Fire Training Center. He had led several training burns when he decided to go to rehab. While at rehab, his vitals were determined to be within normal ranges. He sat in front of a cooling fan and then proceeded to an area near his vehicle to relax.

Moments later, he was found down, suffering from a medical emergency. He was transported to Monongahela Valley Hospital where he passed away. His death was caused by a heart attack.

August 14, 2008–0630 hrs
Tony McGough, Firefighter
Age 44, Volunteer
Amity Fire Department, Arkansas

Firefighter McGough was responding to an emergency medical incident in his POV. He was involved in a single vehicle crash and received fatal injuries.

August 17, 2008–Time Unknown
Robert Arnold Hales, Firefighter
Age 40, Volunteer
Scappoose Rural Fire District, Oregon

Firefighter Hales had recently returned home after fighting lightning-caused wildland fires for 12 hours. He and his daughter left their home in his personal vehicle.

Firefighter Hales' vehicle left the roadway and crashed. His 14-year-old daughter ran a half mile back home and called 9-1-1. Firefighters responded but Firefighter Hales could not be revived. His death was caused by a heart attack.

August 21, 2008–1045 hrs
Curtis Lloyd Jessen, Assistant District Forester
Age 32, Wildland Full-Time
North Carolina Division of Forest Resources, North Carolina

On August 20, 2008, personnel from the North Carolina Division of Forest Resources were dispatched to a wildland fire. Firefighters found a slow moving fire in an area with extremely rough terrain and dangerous rock cliffs. With nightfall approaching, firefighters constructed a containment line around the fire to hold the fire until the next morning.

On the morning of August 21, 2008, Assistant District Forester Jessen joined other firefighters in preparations to mop up the fire. Assistant District Forester Jessen advised the IC that he was going to scout around the other side of the fire. Approximately 15 minutes later, the IC unsuccessfully attempted to contact Assistant District Forester Jessen by radio and cell phones.

Firefighters were sent to check on the safety of Assistant District Forester Jessen. He was found at the bottom of a 66-foot ravine. When firefighters reached Assistant District Forester Jessen, he was deceased. A complicated recovery process was executed by firefighters to remove him from the ravine.

August 25, 2008–Time Unknown
Curtis Ray Hillman, Sr., Firefighter
Age 76, Wildland Contract

TNT Construction under contract to the USFS–Shasta-Trinity National Forest, California

Hillman, a member of the Karuk tribe, was operating a grader to improve road conditions and access for firefighters. They were working the Siskiyou and Blue 2 Complex of fires when Hillman was injured on August 25, 2008.

Heavy Equipment Operator Hillman was working on forest roads 14 and 21, about half a mile from Highway 96 just south of Dillon Creek Campground. The area is halfway between Happy Camp and Orleans.

When his grader failed to start, Hillman and another worker tried to fix the problem. The grader then started, but its brake failed and it began to roll backward. Both men fell or jumped off the machine, and Hillman hit his head.

He was flown to Mercy Medical Center in Redding, where he died from his injuries on September 11, 2008.

September 1, 2008–1810 hrs
Calvin Gene Wahlstrom, Chief Pilot
Age 61, Wildland Contract

Neptune Aviation Services, Inc., Montana

Gregory Jess Gonsioroski, First Officer
Age 41, Wildland Contract

Neptune Aviation Services, Inc., Montana

Zachary Jake Vander-Griend, Crew Chief
Age 25, Wildland Contract

Neptune Aviation Services, Inc., Montana

The California Department of Forestry and Fire Protection contracts with private companies such as Neptune Aviation Services to provide airtanker services. The aircraft involved in this incident was a Lockheed SP-2H airplane that was built in 1962.

The crew was dispatched to provide a water drop on a wildland fire burning in Calaveras County, California. The flight originated at the Reno/Stead (4SD) Airport, in Reno, Nevada.

Continued on next page

An air tanker base employee who witnessed the accident reported observing the airplane taxi to Runway 32 "...and everything appeared normal." The witness reported watching the airplane takeoff, and at an elevation estimated to be between 100 to 300 feet above the ground, he observed the left jet engine emitting flames, followed by the left wing being engulfed in flames. The witness further reported that about 2 seconds later the airplane entered a left wing down attitude before impacting terrain and bursting into flames.

Upon investigation by the NTSB and the Federal Aviation Administration (FAA), pieces of the jet engine were found about 500 feet from the departure end of Runway 32.

For additional information about this crash, consult the NTSB Web site at (http://www.ntsb.gov/ntsb/query.asp - NTSB identification SEA08GA194).

September 3, 2008–1745 hrs
Gregory Allen Northup, Firefighter
Age 55, Volunteer
Gallipolis Fire Department, Ohio

Firefighter Northup was working a special detail for the fire department and was engaged in watering down a house that the fire department was demolishing for the city. At approximately 1745 hours, he walked into the offices of the local newspaper and collapsed.

A newspaper employee started CPR and emergency workers were called. He was taken to the hospital but later died as the result of a heart attack.

September 23, 2008–1600 hrs
Edward A. Junginger, Fire Police Officer–Honorary Chief
Age 82, Volunteer
Levittown Fire Department, New York

Fire Police Officer Junginger was helping to reopen a local highway after a previous automobile crash. He was moving a car driven to the scene by the mother of one of the people involved in the crash when he became ill. He was discovered slumped over the wheel of the car when the woman returned to retrieve her vehicle. The woman started CPR and called for help.

Nearby firefighters and law enforcement officers took over treatment. He was rushed to a local hospital where he was pronounced dead at 1700 hours. His death was caused by a heart attack.

September 29, 2008–1345 hrs
Ralph Paul Arabie, Firefighter
Age 48, Career
David Crockett Steam Fire Co., No. 1, Gretna, Louisiana

Firefighter Arabie was assisting an insurance adjuster as they used a fire department aerial device to assess damage that the fire station had sustained in recent storms.

As he bedded the aerial device, Firefighter Arabie was crushed between the apparatus and the aerial device receiving a fatal head injury.

September 29, 2008–Time Unknown
Dale Wayne Grider, Firefighter
Age 36, Volunteer
DeKalb Fire Department, Texas

Firefighter Grider responded with the members of his fire department to a grass fire. Approximately 4 hours later, Firefighter Grider was riding in a fire department vehicle to training when he became ill. He died of an apparent heart attack.

October 1, 2008–1400 hrs
Douglas Falconer, Correctional Officer II
Age 46, Wildland Part-Time
Arizona State Prison-Globe, Arizona

Officer Falconer had been dispatched on September 30, 2008, with an inmate crew to the Sacramento Fire in the vicinity of Lake Havasu City near the California border with Arizona. The crew had just set up its gear and was working the fire line when Officer Falconer told others that he was dizzy. He was sitting down when he collapsed. He was treated by medical personnel assigned to the fire and transported to a local hospital. He was pronounced dead of a heart attack at 1445 hours.

October 13, 2008–2130 hrs
William J. "Bill" Miller, Firefighter
Age 24, Volunteer
Blue Mound Fire Department, Illinois

Firefighter Miller was driving a tanker (tender) as a part of a training exercise. The tanker was involved in a crash and flipped over. Firefighter Miller's death was caused by a pre-existing medical condition while involved in the rollover. The nature of the pre-existing condition was not known.

A firefighter riding as a passenger in the tanker received minor injuries.

October 16, 2008–Time Unknown
Brian Dennis Neville, Paramedic Firefighter
Age 32, Career

Baltimore County Fire Department, Maryland

Paramedic Firefighter Neville was discovered deceased by other firefighters in his bed on the morning of October 16, 2008. He had responded to several emergency incidents during his shift.

October 29, 2008–1300 hrs
Adam Cody Renfroe, Firefighter
Age 24, Volunteer

Crossville Fire Department, Alabama

Firefighter Renfroe and the members of his fire department were dispatched to a fire in a single-family residence. Firefighter Renfroe was on the first engine to arrive on the scene.

He and another firefighter advanced a hoseline to the front door of the residence. Firefighter Renfroe sent the other firefighter back to the fire truck for a tool. When the firefighter returned, Firefighter Renfroe was gone and the nozzle remained by the doorway. At about the same time, the fire inside of the structure intensified.

Firefighter Renfroe transmitted a distress message from the interior. Firefighters were not immediately able to enter the structure due to fire conditions. By the time firefighters reached him, he was deceased.

Firefighters discovered Firefighter Renfroe about 4 feet from the home's back door. The cause of death was smoke inhalation and thermal burns.

November 3, 2008–0334 hrs
Wayne D. Brown, Firefighter/Fire Police Officer
Age 63, Volunteer

Bristol Fire and Rescue Department, Rhode Island

Firefighter Brown rose from bed and was preparing to respond to a vehicle fire. He became ill and called for responders to assist him. When emergency personnel arrived, they found Firefighter Brown unresponsive, near the door to his home, wearing his fire department equipment.

Firefighter Brown was transported to the hospital where he later died. His death was caused by a heart attack. The vehicle fire was intentionally set and related to a suicide.

November 5, 2008–1702 hrs
Leonard Riggins, Captain
Age 52, Career
Saint Louis Fire Department, Missouri

Captain Riggins was on his way home in uniform and in his fire department vehicle when he happened upon what appeared to be a vehicle crash. According to fire department policy, he pulled to the side of the road, activated his emergency lights, and exited the vehicle to assist. As he approached the vehicle, the occupant of the crashed vehicle turned and shot Captain Riggins in the chest. The occupant had carjacked the vehicle that he crashed. After shooting Captain Riggins, the individual stole Captain Riggins' fire department vehicle and fled the scene.

Assistance was requested for Captain Riggins. He was transported to the hospital where he succumbed to his injuries. The suspect eventually crashed Riggins' vehicle and was attempting to carjack a third vehicle when police approached him and he shot at them. Police returned fire, killing the suspect.

November 7, 2008–0100 hrs
Roy Dale Smith, III, Firefighter
Age 17, Volunteer
McGaheysville Volunteer Fire Department, Virginia

The McGaheysville Volunteer Fire Department responded to an automatic fire alarm at approximately 0047 hours. The first-arriving units found fire coming from the building and upgraded to a working incident. Firefighter Smith responded from his residence in his personal vehicle.

During the response, Firefighter Smith's vehicle left the right side of the road in a curve then crossed the roadway and exited the left side of the roadway. It crashed into a power box, struck some trees, and rolled several times. Firefighter Smith was not wearing a seatbelt at the time of the crash and he was ejected from the vehicle. The vehicle ended up off the road and could not be seen from the road due to the topography and extremely foggy conditions.

Power company representatives searching for the source of an associated power outage discovered the crash scene and reported it at approximately 0325 hours. Fire and rescue units responded to the crash scene and discovered Firefighter Smith. He was pronounced dead at the scene.

The bylaws of the McGaheysville Volunteer Fire Department permit full membership in the department as a firefighter at age 17.

November 9, 2008–1500 hrs
Cecilia Turnbough, EMT/Firefighter Recruit
Age 42, Volunteer

Dale City Volunteer Fire Department, Virginia

From the Prince William County report on the incident:

A Volunteer Firefighter I training program was being conducted at the Prince William County Public Safety Training Center. On this day, the students were being instructed on the proper techniques to overcome obstacles and entry into the maze while using full PPE and SCBA. The components for this class consisted of classroom training, PPE dressing drills, hose maze, SCBA skill station/low profile, and a maze practical exercise.

Firefighter Recruit Turnbough was a student in this Volunteer Firefighter I class and had completed all components leading up to the maze practical. Firefighter Recruit Turnbough started breathing air from her SCBA at 1414 hours and entered the maze at approximately 1417 hours with 4,260 pounds per square inch of air in her SCBA cylinder. While moving through the maze, the instructors remained in voice contact with her up to the point where she entered the incline/decline section of the maze.

At approximately 1438 hours, the instructors realized Firefighter Recruit Turnbough was having some problems in the maze. They went into the incline/decline section of the maze to remove her. According to the information from a computer monitor integrated with her SCBA, she stopped breathing air from her SCBA at 1438 hours and her PASS alarm activated due to lack of motion at 1439 hours.

When the instructors reached Firefighter Recruit Turnbough, she did not respond to verbal commands and was unresponsive. The two instructors inside of the maze attempted to remove her prior to notifying the other instructors of the emergency. When the third instructor entered the maze, he was informed of the emergency and was directed to call 9-1-1. The fourth instructor dialed 9-1-1 at 14:45:35 hours. Firefighter Recruit Turnbough was removed from the maze at approximately 1449 hours and CPR was started by the instructors.

Medic 525 arrived on the scene and started advanced life support (ALS) care. She was transported to Prince William Hospital where resuscitation efforts continued. Firefighter Recruit Turnbough was pronounced dead at 1537 hours. The cause of death was cardiac-related; the manner of death was listed as natural.

November 10, 2008–1130 hrs
Jamel M. Sears, Probationary Firefighter
Age 33, Career

Fire Department City of New York, New York

Probationary Firefighter Sears joined the FDNY in July, 2008, as part of a 23-week probationary firefighter training program. After completing an 18-minute training evolution at the department's training academy in Randall's Island, Sears collapsed and was unresponsive. Firefighters at the scene attempted to revive him. He was transported to the Mount Sinai Medical Center, where he passed away November 11, 2008. The cause of death was listed as acidosis and dehydration.

November 11, 2008–1440 hrs
Eugene William Franklin, Jr., Chaplain
Age 64, Volunteer

Sumter Fire Department, South Carolina

Chaplain Franklin was driving from his residence to a local hospital in Sumter to console members of the Sumter Fire Department. He was visiting a member whose father was near death and a retired Engineer who was hospitalized.

As he drove to the hospital on his motorcycle, a vehicle turned in front of him and he was killed as a result of the collision.

November 14, 2008–Time Unknown
Carol Irene Taylor, Firefighter
Age 41, Career

Goldsboro Fire Department, North Carolina

Firefighter Taylor had gotten off shift at 0800 hours on November 14, 2008. She had responded to an emergency incident after midnight that morning. She was last seen alive at 1830 hours that evening. After not reporting back to work the morning of November 15, 2008, and not answering her phone, she was found deceased in her home by other firefighters. The cause of death was a heart attack.

November 15, 2008–0530 hours
Walter Patrick Harris, Sr., Senior Firefighter
Age 38, Career

Detroit Fire Department, Michigan

Firefighter Harris responded to an incendiary fire in an abandoned house fire. He and several other firefighters were putting out hotspots in the attic. The roof collapsed trapping the firefighters. Firefighter Harris was crushed by the debris. He was rushed to the hospital but succumbed to his injuries. The other firefighters managed to escape with minor injuries.

November 17, 2008–0838 hrs
Michael David Snowman, Firefighter
Age 49, Volunteer

Hartland Volunteer Fire Department, Maine

Firefighter Snowman responded to the scene of a mutual-aid structure fire in his personal vehicle. When he arrived on the scene, he began to assist with apparatus operations and was stricken with a heart attack as he attached a hoseline to a truck.

Firefighter Snowman was treated on the scene and transported by ambulance to the hospital. He was pronounced dead at 1150 hours.

November 18, 2008–1115 hours
Steven D. "Tiny" Kline, Firefighter/Paramedic
Age 37, Part-Time (Paid)

Stone Park Fire Department, Illinois

Firefighter/Paramedic Kline conducted station duties and participated in classroom training on the morning of November 18. Another firefighter left the room for a moment and when he returned, he found Firefighter Kline unresponsive.

He was transported to the hospital but did not recover. The cause of death was listed as a heart attack.

November 23, 2008–0040 hours
Robert J. Ryan, Jr., Lieutenant
Age 46, Career

Fire Department City of New York, New York

Lieutenant Ryan and the members of his engine company responded to a structure fire in a residence. He and his crew were working in the attic of the structure when a collapse occurred.

Lieutenant Ryan's helmet and facepiece were dislodged by the collapse. He was rescued and brought to the exterior by other firefighters. He was taken to the hospital but did not survive. The cause of death was smoke inhalation.

November 29, 2008–0051 hours
Clarence E. Watson, II, Lieutenant
Age 35, Career

Hematite Fire Protection District, Missouri

Lieutenant Watson was on duty from 0800 hours on November 27 through 0800 hours on November 28. He responded to at least one incident during the shift at 0135 hours. After being relieved, Lieutenant Watson responded to another incident at 1013 hours.

Firefighters and EMS responders were called to Lieutenant Watson's home at 0051 hours on November 29 and found him in full cardiac arrest.

December 8, 2008–2040 hrs
Rick H. Borkin, Lieutenant
Age 42, Volunteer

Thiensville Fire Department, Wisconsin

Lieutenant Borkin was participating in fire department training. During a particularly physically challenging portion of the training, Lieutenant Borkin indicated that he was ill and sat down. Shortly thereafter, he collapsed.

Lieutenant Borkin was treated by firefighters and local paramedics. He was pronounced dead at the scene. The cause of death was a heart attack.

December 17, 2008–1148 hrs
Jerry James Parrick, Firefighter
Age 59, Volunteer

West End Volunteer Fire and Rescue, Montana

Firefighter Parrick and members of his fire department responded to a single vehicle crash on a local highway which was divided by a median. The crash scene was located in a curve. Firefighter Parrick responded to the scene in his personal vehicle, a four-wheel drive pickup truck equipped with emergency lights.

The IC directed Firefighter Parrick to park his personal vehicle upstream of the crash to warn approaching drivers. Firefighter Parrick positioned his vehicle at the side of the road and activated his emergency lights.

A tractor-trailer truck towing tandem trailers lost control and struck Firefighter Parrick's vehicle. Firefighter Parrick was sitting in the driver's seat. He was discovered by other firefighters lying in the rear seat of the pickup. He was pronounced dead at the scene.

The cause of death was listed as blunt force trauma. Firefighter Parrick was most likely killed as the result of striking parts of the interior of the vehicle during the crash.

For additional information regarding this incident, please refer to NIOSH Fire Fighter Fatality Investigation and Prevention Program Report F2009-03 (http://www.cdc.gov/niosh/fire/reports/face200903.html).

December 20, 2008–2152 hrs
Michelle Newton Smith, Firefighter
Age 29, Volunteer

Delaware City Fire Department, Delaware

Firefighter Smith was working on the scene of a motorcycle crash. The driver of a passenger vehicle entering the scene lost control and struck a law enforcement vehicle. Firefighter Smith was struck and severely injured. She was treated at the scene and in the hospital but died of her injuries on December 22, 2008.

The driver of the vehicle that killed Firefighter Smith fled the scene. Law enforcement officials later arrested a man and charged him with first degree murder, first degree assault, driving with a revoked license, leaving the scene of a crash, and failing to report a crash. The driver is a paraplegic and is alleged to have been using a stick to operate the controls of the car at the time of the crash.

December 22, 2008–2330 hrs
Stephen Hagan, Sr., Lieutenant
Age 48, Volunteer

Blenheim Fire Department, South Carolina

Lieutenant Hagan responded to the scene of a motor vehicle crash. While working at the scene, he began to feel sick and was taken to the hospital in Bennettsville. Lieutenant Hagan was subsequently transported by ambulance to the hospital in Florence, South Carolina, where he went into cardiac arrest and was unable to be revived.

December 27, 2008–1625 hrs
Dennis G. McClenahan, Driver
Age 54, Volunteer

Princeton Junction Volunteer Fire Co. #1, New Jersey

Driver McClenahan and the members of his fire department responded to a fire alarm incident at a local mall. The incident was concluded and firefighters returned to their homes. A few hours after arriving home, Driver McClenahan collapsed and subsequently died of a heart attack.

December 29, 2008–1840 hrs
Bret S. Kaneshiro, Firefighter I
Age 37, Career

Honolulu Fire Department, Hawaii

Firefighter Kaneshiro's last work shift began at 0800 hours on December 29, 2008, and ended at 0800 hours on December 30, 2008. Firefighter Kaneshiro performed station and apparatus maintenance, attended a morning meeting, and conducted fireworks inspections. His company was dispatched to a medical alarm at 1138 hours but was cancelled while en route.

At approximately 1830 hours on December 29, 2008, Firefighter Kaneshiro participated in a physical fitness training activity which consisted of pull-ups, sit-ups, push-ups, and rowing machine for maximum repetitions in 1 minute. At 2001 hours, Firefighter Kaneshiro's company was dispatched to a medical alarm involving a female victim suffering from a severe nose bleed, who also weighed approximately 250 pounds.

The victim was stabilized and placed on a gurney for transport via ambulance to a hospital. Firefighter Kaneshiro took the rear position on the gurney and with another firefighter at the front position. He struggled to navigate through the tight spaces of the residence, which was cluttered with household items. The company returned to quarters and completed apparatus maintenance at approximately 2100 hours.

On December 30, 2008, at 0800 hours, Firefighter Kaneshiro completed his shift. At approximately 0930 hours, he reported to his part-time job where he spent the majority of his time in the office assisting project managers with computer-related tasks.

Firefighter Kaneshiro then proceeded to the Saint Louis High School soccer field and participated in soccer practice as the goalie coach. According to witnesses, at approximately 1730 hours, Firefighter Kaneshiro felt tired, sat down, and then collapsed. CPR was performed by an offduty firefighter, who found Firefighter Kaneshiro pulseless and instructed the athletic trainer to start ventilations while he applied the school's AED.

Firefighter Kaneshiro was transported to the hospital where he was pronounced dead at 1840 hours.

December 30, 2008–1245 hrs
Norman C. Koch, Fire Police Captain
Age 79, Volunteer

East Pembroke Fire Department, New York

Fire Police Captain Koch and members of his fire department responded to a motor vehicle crash in their community. Fire Police Captain Koch assisted with scene safety and traffic control on the scene. As firefighters completed their work and began to clear the scene, Fire Police Captain Koch collapsed of an apparent heart attack. He was transported to the hospital but did not survive.

December 31, 2008–1759 hrs
Richard Lee Montgomery, Captain
Age 54, Volunteer

Pisgah Community Volunteer Fire Department, Mississippi

Captain Montgomery was assisting with a fire fight in a residential structure. Captain Montgomery exited the structure and suddenly collapsed. His death was caused by a heart attack.

December 31, 2008–2110 hrs
Jarrett Lee Little, Firefighter
Age 24, Paid-on-Call

Walker County Fire and Rescue, Georgia

Firefighter Little was the driver of an engine/tanker apparatus responding to an emergency incident. The engine was equipped with a 1,500-gallon water tank.

As the apparatus made a left-hand turn, Firefighter Little lost control of the vehicle. The engine struck a telephone pole and overturned. Firefighter Little was removed from the vehicle and flown by medical helicopter to a hospital where he was later pronounced dead.

Two other firefighters who were occupants of the apparatus were injured.

Firefighter Fatalities from Previous Years

June 29, 1999–1346 hrs
John Clasby, Firefighter
Age 45, Career

Hull Fire Department, Massachusetts

Firefighter Clasby and the members of his fire department were fighting a wind-driven fire that involved four houses. As Firefighter Clasby and another firefighter worked in the back yard of a home, several explosions were heard. Firefighter Clasby fell to the ground.

When he was treated at the hospital, it was discovered that a bullet had severed his spinal cord and left him paralyzed. Firefighter Clasby died as a result of infection complications from his injury on November 11, 2008.

September 11, 2001
Glenn J. Winuk, Firefighter
Age 40, Volunteer

Jericho Fire Department, New York

Glenn J. Winuk served the Jericho Fire District as a volunteer firefighter, emergency medical technician (EMT), lieutenant, and Fire Commissioner over the course of a 20-year career.

On September 11, 2001, Mr. Winuk, a partner at the law firm Holland & Knight LLP, raced from his nearby evacuated office to participate in the rescue effort at the World Trade Center. He died when the South Tower collapsed, medic bag by his side and protective gloves on his hands.

Appendix B

Firefighter Fatality Inclusion Criteria—National Fire Service Organizations

The National Fire Protection Association (NFPA), the National Fallen Firefighters Foundation (NFFF), the United States Fire Administration (USFA), and other organizations individually collect information on firefighter fatalities in the United States. Each organization uses a slightly different set of inclusion criteria that are based at least in part on the purposes of the information collection for each organization and data consistency.

As a result of these differing inclusion criteria, statistics about firefighter fatalities may be provided by each organization that do not coincide with one another. This section will explain the inclusion criteria for each organization and provide information about these differences.

Firefighter Fatalities in 2008
Incidents Occuring in 2008

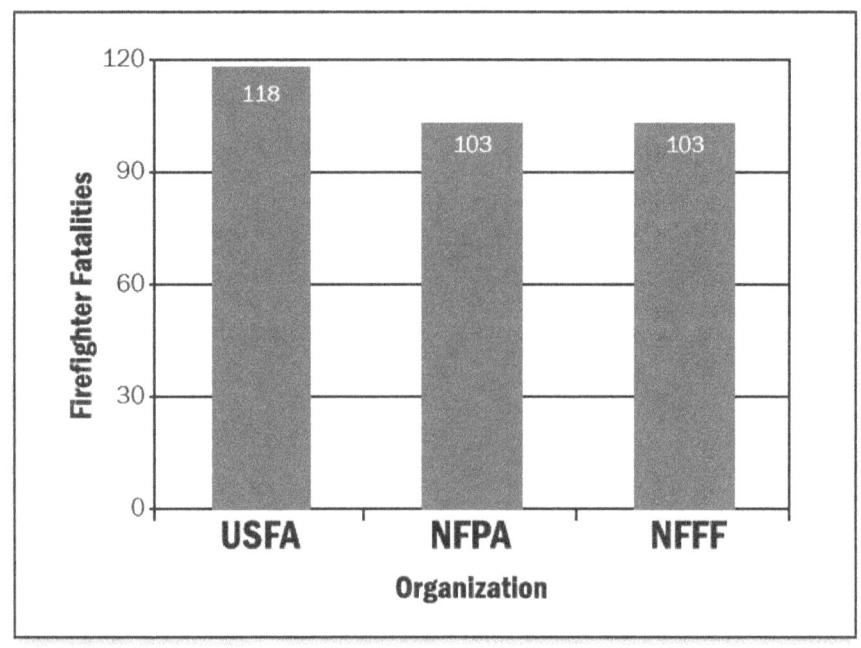

The USFA includes firefighters in this report who died while on duty, became ill while on duty and later died, and firefighters who died within 24-hours of an emergency response or training regardless of whether the firefighter complained of illness while on duty. The USFA counts firefighter deaths that occur in the 50 States, the District of Columbia, and United States protectorates such as Puerto Rico and Guam.

For 2008, USFA reported 118 onduty firefighter fatalities.

Inclusion Criteria for the National Fire Protection Association's Annual Firefighter Fatality Study

Introduction

Each year, the National Fire Protection Association (NFPA) collects data on all firefighter fatalities in the United States that resulted from injuries or illnesses that occurred while the victims were on duty. The purpose of the study is to analyze trends in the types of illnesses and injuries resulting in death that occur while firefighters are on the job. This annual census of firefighter fatalities in its current format dates back to 1977. (Between 1974 and 1976, NFPA published a study of onduty firefighter fatalities that was not as comprehensive.)

What is a firefighter?

For the purpose of the NFPA study, the term *firefighter* covers all uniformed members of organized fire departments, whether career, volunteer, combination, or contract; full-time public service officers acting as firefighters; State and Federal government fire service personnel; temporary fire suppression personnel operating under official auspices of one of the above; and privately-employed firefighters including trained members of industrial or institutional fire brigades, whether full- or part-time.

Under this definition, the study includes, besides uniformed members of local career and volunteer fire departments, those seasonal and full-time employees of State and Federal agencies who have fire suppression responsibilities as part of their job description, prison inmates serving on firefighting crews, military personnel performing assigned fire suppression activities, civilian firefighters working at military installations, and members of industrial fire brigades. Impressed civilians would also be included if called on by the officer in charge of the incident to carry out specific duties. The NFPA study includes fatalities that occur in the 50 States and the District of Columbia.

What does "on duty" mean?

The term *on duty* refers to being at the scene of an alarm, whether a fire or nonfire incident; being en route while responding to or returning from an alarm; performing other assigned duties such as training, maintenance, public education, inspection, investigations, court testimony and fundraising; and being on-call, under orders or on standby duty other than at home or at the individual's place of business. Fatalities that occur at a firefighter's home may be counted if the actions of the firefighter at the time of injury involved firefighting or rescue.

Onduty fatalities include any injury sustained in the line of duty that proves fatal, any illness that was incurred as a result of actions while onduty that proves fatal, and fatal mishaps involving nonemergency occupational hazards that occur while on duty. The types of injuries included in the first category are mainly those that occur at an incident scene, in training, or in accidents while responding to or returning from alarms. Illnesses (including heart attacks) are included when the exposure or onset of symptoms are tied to a specific incident of onduty activity. Those symptoms must have been in evidence while the victim was on duty for the fatality to be included in the study.

Fatal injuries and illnesses are included even in cases where death is considerably delayed. When the onset of the condition and the death occur in different years, the incident is counted in the year of the condition's onset. Medical documentation specifically tying the death to the specific injury is required for inclusion of these cases in the study.

Categories not included in the study

The NFPA study does not include members of fire department auxiliaries; nonuniformed employees of fire departments; emergency medical technicians (EMTs) who are not also firefighters; chaplains; or civilian dispatchers. The study also does not include suicides as onduty fatalities even when the suicide occurs on fire department property.

The NFPA recognizes that a comprehensive study of firefighter onduty fatalities would include chronic illnesses (such as cardiovascular disease and certain cancers) that prove fatal and that arose from occupational factors. In practice, currently there is no mechanism for identifying onduty fatalities that are due to illnesses that develop over long periods of time. This creates an incomplete picture when comparing occupational illnesses to other factors as causes of firefighter deaths. This is recognized as a gap the size of which cannot be identified at this time because of the limitations in tracking the exposure of firefighters to toxic environments and substances and the potential long-term effects of such exposures.

2008 Experience

In 2008, according to the NFPA inclusion criteria a total of 103 onduty firefighter deaths occurred in the United States.

National Fallen Firefighters Foundation

In 1997, fire service leaders formulated new criteria to determine eligibility for inclusion on the National Fallen Firefighter Memorial. Line-of-duty deaths (LODDs) shall be determined by the following standards:

1. (a) Deaths of firefighters meeting the Department of Justice's Public Safety Officers' Benefits (PSOB) program guidelines, and those cases that appear to meet these guidelines whether or not PSOB staff has adjudicated the specific case prior to the annual National Fallen Firefighters Memorial Service; and

    (b) Deaths of firefighters from injuries, heart attacks or illnesses documented to show a direct link to a specific emergency incident or department-mandated training activity.

2. While PSOB guidelines cover only public safety officers, the Foundation's criteria also include contract firefighters and firefighters employed by a private company, such as those in an industrial brigade, provided that the deaths meet the standards listed above.

3. Some specific cases will be excluded from consideration, such as deaths attributable to suicide, alcohol or substance abuse, or other gross abuses as specified in the PSOB guidelines.

The National Fallen Firefighters Memorial was built in 1981 in Emmitsburg, Maryland. The names listed there begin with those firefighters who died in the line-of-duty that year. The U.S. Congress created the National Fallen Firefighters Foundation to lead a nationwide effort to remember America's fallen firefighters. Since 1992, the tax-exempt, nonprofit Foundation has developed and expanded programs to honor our fallen fire heroes and assist their families and coworkers by providing them with resources to rebuild their lives. Since 1997, the Foundation has managed the National Memorial Service held each October to honor the firefighters who died in the line of duty the previous year.

At the October 2009 Memorial Weekend, the Foundation will be honoring 122 firefighters who died in the line-of-duty. Of those 122 being honored, 103 died in 2008 as the result of incidents that occurred in 2008, 5 firefighters died in 2007 as the result of an incident that occurred in 2007, and 14 others died in previous years as the result of incidents that occurred in previous years. For a listing of the firefighters that will be honored by the Foundation in October of 2009, please visit their website at www.firehero.org.

www.ingramcontent.com/pod-product-compliance
Lightning Source LLC
Chambersburg PA
CBHW081143170526
45165CB00008B/2772

9781482764161